An Introduction to FUNCTIONAL PROGRAMMING Through LAMBDA CALCULUS

Greg Michaelson
Heriot-Watt University

DOVER PUBLICATIONS, INC.
Mineola, New York

Copyright

Bibliographical Note

This Dover edition, first published in 2011, is an unabridged republication of the work originally published in 1989 by Addison-Wesley Publishing Company, Wokingham, England. The author has provided a new Preface for this edition.

Library of Congress Cataloging-in-Publication Data

Michaelson, Greg, 1953–
An introduction to functional programming through Lambda calculus / Greg Michaelson. — Dover ed.
p. cm.
Originally published: Workingham, England : Addison-Wesley, 1989.
Includes bibliographical references and index.
ISBN-13: 978-0-486-47883-8
ISBN-10: 0-486-47883-1
1. Functional programming (Computer science). 2. Lambda calculus. I. Title.

QA76.6.M4854 2011
005.1'14—dc22

2010031017

Manufactured in the United States by LSC Communications
47883104 2016
www.doverpublications.com

Preface to the Dover Edition, 2011

Context

When I started to write this book in 1986, functional programming seemed on an upward trajectory, out of academia into "real-world" computing. Spurred by the Japanese 5th Generation programme, many other nations initiated research and development schemes around "stateless" declarative programming languages. The Japanese projects laid great stress on hardware and software to support logic programming. In contrast, the UK Alvey programme gave equal emphasis to functional languages. Thus, major British multi-University and industry projects sought to develop viable functional computing platforms at all levels, from VLSI and multi-processor hardware for graph reduction and combinators, like Cobweb, Alice and GRIP, through novel programming languages, like Hope, Standard ML and Haskell, to fully integrated systems like the Flagship.

Despite lasting research successes, overall the 5th Generation programmes utterly failed to make any impact on the von Neumann/imperative language status quo, and are now largely forgotten. In particular, declarative languages were oversold as the solution to the "software crisis". The core claim was that the absence of state provided a model that greatly simplified high levels of abstraction, component based software construction, parallel implementation and proof of correctness. My original Introduction to this book is one such example of zeitgeist enthusiasm.

While declarative languages have many advantages for problems which are highly compositional in their data, the lack of state is a major hindrance for most substantive applications which do things *in* rather than *beyond* time. Similarly, while proving properties of declarative programs may, for small exemplars, be somewhat easier than for imperative code, this again depends on highly compositional program structure. The bottom line is that constructing and proving correct programs is hard, be it in a declarative or imperative language.

Nonetheless, research in functional programming has remained buoyant and its industrial significance is steadily growing, if somewhat more slowly than anticipated 25 years ago. Most strikingly, Ericsson's Erlang, originally used for telecommunications, is gaining wider currency for general multi-process systems. Furthermore, Microsoft Research supports the development of the Haskell language and has integrated its own Standard ML-derivedF# into the core .NET framework. Constructs from functional languages have also found their way into mainstream imperative languages. Thus, functional abstraction is supported in Microsoft's system language C#, the general purpose Ruby, the scripting language Python and the Java-derived Groovy. Finally, cheap high performance computing on tens of thousands of commodity platforms has enabled international Internet businesses to build services on standardised parallel frameworks, based explicitly on higher order constructs. For example, Google's search engines are driven by their proprietary Map-Reduce, and Yahoo deploys Apache's open source Hadoop.

And, 75 years on from Alonzo Church's pioneering exploration of the entscheidungsproblem[1], λ calculus remains central to Computing education and research. Functional programming is a routine component of many Computer Science undergraduate programmes, as are computability theory, and, to a lesser extent, formal semantics. Functional techniques are being used in novel approaches to constructing scalable heterogeneous multi-processor systems and to developing mission and safety critical systems requiring strong assurances that requirements are met. New calculi building on l calculus have been systematically developed for modelling multi-process systems, for example Milner's Communicating Concurrent Systems, ∏ Calculus and bigraphs.

Content

Looking at this book from a markedly older and greyer perspective, by and large I feel happy with it. In particular, I remain firmly wedded to the pedagogy of learning by abstraction from concrete examples, of understanding λ calculus through actually "doing" it in an explicitly operational manner, and of gaining oversight of the layers between a simple, foundational system and a rich language of variegated constructs and structures.

The book's major eccentricity remains my reformulation of classic λ calculus syntax. In Church's notation, applications in function bodies are unbracketed, but functions are bracketed except where there is no ambiguity. I chose instead to bracket applications in function bodies but to not bracket functions except where there is ambiguity, as I felt these were more in keeping with programming language conventions. In retrospect, I suspect that this may prove unduly confusing to novices trying to use this book to complement other sources.

I now think that the account of lazy evaluation could be simplified. There is also merit in one reviewer's suggestion[2] that a pure lazy polymorphic language might have been described given that both Lisp and SML are strict and impure. However, when the book was written, Miranda™ was a commercial product and Haskell had not been standardised.

Finally, my favourite chapter remains that on recursion.

Conclusion

When I was wee, my parents had lots of Dover books: Charles Babbage, Lewis Carroll, Gustave Dore ... I am tickled pink to be in the company of such authors.

So, I would very much like to thank:

- John Crossley for suggesting that I approach Dover;
- John Grafton at Dover for reprinting this book, and for all his support.

Greg Michaelson, Edinburgh, May 2011.

[1] Church, Alonzo, *An unsolvable problem of elementary number theory*, American Journal of Mathematics, 58 (1936), pp. 345–363.

[2] R. Jones, Times Higher Education Supplement, p29, 29/9/89.

Preface

Overview

This book aims to provide a gentle introduction to functional programming. It is based on the premise that functional programming provides pedagogic insights into many aspects of computing and offers practical techniques for general problem solving.

The approach taken is to start with pure λ calculus, Alonzo Church's elegant but simple formalism for computation, and to add syntactic layers for function definitions, booleans, integers, recursion, types, characters, lists and strings to build a fairly high level functional notation. Along the way, a variety of topics are discussed including arithmetic, linear list and binary tree processing, and alternative evaluation strategies. Finally, functional programming in Standard ML and Common LISP, using techniques developed throughout the book, are explored.

Readership

This book is intended for people who have taken a first course in an imperative programming language like Pascal, FORTRAN or C and have written programs using arrays and subprograms. There are no mathematical prerequisites and no prior experience with functional programming is required.

The material from this book has been taught to third year undergraduate Computer Science students and to postgraduate Knowledge-Based Systems MSc students.

Using the book

The material is presented sequentially. Each chapter depends on previous chapters. Within chapters, substantial use is made of worked examples. Each chapter ends with exercises which are based directly on ideas and techniques from that chapter. Specimen answers are included at the end of the book.

Approach

Within this book, λ calculus is the primary vehicle for developing functional programming. I was trained in a tradition which saw λ calculus as a solid base for understanding computing and my own teaching experience confirms this. Many books on functional programming cover λ calculus but the presentation tends to be relatively brief and theoretically oriented. In my experience, students whose first language is imperative find functions, substitution and recursion conceptually difficult. Consequently, I have given a fair amount of space to a relatively informal treatment of these topics and include many worked examples. Functional programming aficionados may find this somewhat tedious. However, this is an introductory text.

This book does not try to present functional programming as a complete paradigm for computing. Thus, there is no material on the formal semantics of functional languages or on transformation and implementation techniques. These topics are ably covered in other books. By analogy, one does not buy a book on COBOL programming in anticipation of chapters on COBOL's denotational semantics or on how to write COBOL compilers. However, a number of topics which might deserve more thorough treatment are omitted or skimmed. In particular, there might be more discussion of types and typing schemes, especially abstract data types and polymorphic typing, which are barely mentioned here. I feel that these really deserve a book to themselves but hope that their coverage is adequate for what is primarily an introductory text. There is no mention of mutual recursion which is conceptually simple but technically rather awkward to present. Finally, there is no discussion of assignment in a functional context.

The functional notation developed in the book does not correspond to any one implemented language. One of the book's objectives is to explore different approaches within functional programming and no single language encompasses these. In particular, no language offers different reduction strategies.

The final chapters consider functional programming in Standard ML and Common LISP. Standard ML is a modern functional language with succinct syntax and semantics based on sound theoretical principles. It is a pleasing language in which to program and its use is increasing within education and research. SML's main pedagogic disadvantage is that it lacks normal order reduction and so the low-level λ calculus representations discussed in earlier chapters cannot be fully investigated in this language.

LISP was one of the earliest languages with an approximation to a functional subset. It has a significant, loyal following, particularly in the artificial intelligence community, and is programmed using many functional techniques. Here, Common LISP was chosen as a widely used modern LISP. Like SML, it lacks normal order reduction. Unlike SML, it combines minimal syntax with baroque semantics, having grown piecemeal since the late 1950s.

About the chapters

In Chapter 1, we will look at the differences between imperative and functional programming, consider the origins of functional programming in the theory of computing, survey its role in contemporary computing and discuss λ calculus as a basis for functional programming.

In Chapter 2, we will look at pure λ calculus, examine its syntax and evaluation rules, and develop functions for representing pairs of objects. These will be used as building blocks in subsequent chapters. We will also introduce simplified notations for λ expressions and for function definitions.

In Chapter 3, we will develop representations for boolean values and operations, numbers and conditional expressions.

In Chapter 4, we will develop representations for recursive functions and use them to construct arithmetic operations.

In Chapter 5, we will discuss types and introduce typed representations for boolean values, numbers and characters. Notations for case definitions of functions will also be introduced.

In Chapter 6, we will develop representations for lists and look at linear list processing.

In Chapter 7, linear list processing techniques will be extended to composite values and we will consider nested structures such as trees.

In Chapter 8, we will discuss different evaluation orders and termination.

In Chapter 9, we will look at functional programming in Standard ML, making use of the techniques developed in earlier chapters.

In Chapter 10, we will look at functional programming in LISP using the techniques developed in earlier chapters.

Notations

In this book, different typefaces are used for different purposes. Text is in Times Roman. **New terms and important concepts are in Times Bold.** Programs and definitions are in Helvetica. Greek characters are used in naming λ calculus concepts:

α – alpha
β – beta
λ – lambda
η – eta

Syntactic constructs are defined using Backus-Naur-Form (BNF) **rules**. Each rule has a **rule name** consisting of one or more words within angle brackets < and >. A rule associates its name with a **rule body** consisting of a sequence of **symbols** and rule names. If there are different possible rule bodies for the same rule then they are separated by '|'. For example, binary numbers are based on the digits 1 and 0:

```
<digit> ::= 1 | 0
```

and a binary number may be either a single digit or a digit followed by a number:

```
<binary> ::= <digit> | <digit> <binary>
```

Acknowledgements

I had the good fortune to be taught Computer Science at the University of Essex from 1970 to 1973. There I attended courses on the theory of computing with Mike Brady and John Laski, which covered λ calculus, recursive function theory and LISP, and on programming languages with Tony Brooker, which also covered LISP. Subsequently, I was a postgraduate student at St Andrews University from 1974 to 1977 where I learnt about functional language design and implementation from Tony Davie and Dave Turner. I would like to thank all these people for an excellent education.

I would also like to thank my colleagues at Napier College, Glasgow University and Heriot-Watt University with whom I have argued about many of the ideas in this book, in particular Ken Barclay, Bill Findlay, John Patterson, David Watt and Stuart Anderson.

I would, of course, like to thank everyone who has helped directly with this book:

- Paul Chisholm for patiently and thoroughly checking much of the material: his help has been invaluable.
- David Marwick for checking an early draft of Chapter 1 and Graeme Ritchie for checking an early draft of Chapter 10.
- Peter King, Chris Miller, Donald Pattie, Ian Crorie and Patrick McAndrew, in the Department of Computer Science, Heriot-Watt University, who provided and maintained the UNIX facilities used to prepare this book.
- Bob Colomb at CSIRO Division of Information Technology, Sydney for providing a most pleasant environment within which to complete this book.

- Mike Parkinson and Stephen Troth at Addison-Wesley for their help in the development of this book, and Andrew McGettrick and Jan van Leeuwen for their editorial guidance.

  I would particularly like to thank Allison King at Addison-Wesley.

  Finally, I would like to thank my students.

I alone am responsible for errors lurking within these pages. If you spot any then please let me know.

Greg Michaelson

Edinburgh and Sydney
1988

Contents

Chapter 1
Introduction

Functional programming is an approach to programming based on function calls as the primary programming construct. It provides practical approaches to problem solving in general and insights into many aspects of computing. In particular, with its roots in the theory of computing, it forms a bridge between formal methods in computing and their application. In this chapter we are going to look at how functional programming differs from traditional imperative programming. We will do this by directly contrasting the imperative and functional approaches to various aspects of programming. We will then consider the origins of functional programming in the theory of computing and survey its relevance to contemporary computing theory and practice. Finally, we will discuss the role of lambda (λ) calculus as a basis for functional programming.

1.1 Names and values in programming

We write computer programs to implement solutions to problems. First, we analyse the problem. Then, we design a solution and implement it using a programming language. Solving a problem involves carrying out operations on values. Different values are used to solve different instances of a problem. If the values for a particular instance were built into the program, then they would have to be changed when the program was used to solve a different instance.

A fruitful approach to problem analysis is to try to identify a general case of the problem. Programming languages enable the implementation of general case solutions through the use of names to stand for arbitrary values. Thus, we write a program using names to stand for values in general. We then run the program with the names taking on particular values from the input for particular instances of the problem. The program does not have to be changed to be used with different values to solve a different instance of the problem: we simply change the inputs and the computer system makes sure that they are used with the right names in the program.

As we will see, the main difference between imperative programming languages, like Pascal, FORTRAN and COBOL, and functional programming languages, like SML and Miranda, lies in the rules governing the association of names and values.

1.2 Names and values in imperative and functional languages

Traditional programming languages are based around the idea of a **variable** as a changeable association between a name and values. These languages are said to be **imperative** because they consist of sequences of **commands**:

```
<command1> ;
<command2> ;
<command3> ;
 ...
```

Typically, each command consists of an **assignment** which changes a variable's value. This involves working out the value of an expression and associating the result with a name:

```
<name> := <expression>
```

In a program, each command's expression may refer to other variables whose values may have been changed by preceding commands. This enables values to be passed from command to command.

Functional languages are based on structured **function calls**. A functional program is an expression consisting of a function call which calls other functions in turn:

```
<function1>(<function2>(<function3> ... ) ... )
```

Thus, each function receives values from and passes new values back to the calling function. This is known as function **composition** or **nesting.**

In imperative languages, commands may change the value associated with a name by a previous command so each name may be and usually will be associated with different values while a program is running.

> **In imperative languages, the same name may be associated with different values.**

In functional languages, names are only introduced as the formal parameters of functions and given values by function calls with actual parameters. Once a formal parameter is associated with an actual parameter value there is no way for it to be associated with a new value. There is no concept of a command which changes the value associated with a name through assignment. Thus, there is no concept of a command sequence or command repetition to enable successive changes to values associated with names.

> **In functional languages, a name is only ever associated with one value.**

1.3 Execution order in imperative and functional languages

In imperative languages, the order in which commands are carried out is usually crucial. Values are passed from command to command by references to common variables and one command may change a variable's value before that variable is used in the next command. Thus, if the order in which commands are carried out is changed then the behaviour of the whole program may change. For example, in the program fragment to swap X and Y:

```
T := X ;
X := Y ;
Y := T
```

T's value depends on X's value, X's value depends on Y's value and Y's value depends on T's value. Thus, any change in the sequence completely changes what happens. For example:

```
X := Y ;
T := X ;
Y := T
```

sets X to Y and:

```
T := X ;
Y := T ;
X := Y
```

sets Y to X.

Of course, not all command sequences have fixed execution orders. In many imperative languages, the order in which expressions are executed may not be defined. Thus, for expressions which involve function calls, the order in which the functions are called may not be defined. Functions have blocks of commands for bodies. Thus, the order in which the different command blocks are executed may not be defined.

This may lead to problems when imperative languages allow **side effects** – changes to variables made by expressions, for example, when a function changes a non-local variable by assignment to one of its parameters or to a global variable. If the order in which subexpressions are evaluated is unpredictable, then the order in which side effects occur is unpredictable. This makes it very hard to understand, develop and debug programs which utilize them.

If commands' expressions do not refer to each other, then the command execution order does not matter. However, programs usually depend on the precise order in which commands are carried out.

In general, imperative languages have fixed command execution orders.

Pure functional languages lack assignment and so the values associated with names never change. Thus, there are no side effects and function calls cannot change the values associated with common names. Hence, the order in which nested function calls are carried out does not matter because function calls cannot interact with each other. For example, suppose we write functions in a style similar to Pascal:

```
FUNCTION F( X,Y,Z:INTEGER):INTEGER ;
BEGIN ... END
FUNCTION A(P:INTEGER):INTEGER ;
```

```
BEGIN ...  END
FUNCTION B(Q:INTEGER):INTEGER ;
BEGIN ...  END
FUNCTION C(R:INTEGER):INTEGER ;
BEGIN ...  END
```

In a functional language, in the function call:

```
F(A(D),B(D),C(D))
```

the order in which A(D), B(D) and C(D) are carried out does not matter because the functions A, B and C cannot change their common actual parameter D.

In functional languages, there is no fixed execution order.

Of course, functional programs must be executed in some order – all programs are – but the order does not affect the final result. As we shall see, this execution order independence is one of the strengths of functional languages and has led to their use in a wide variety of formal and practical applications.

1.4 Repetition in imperative and functional languages

In imperative languages, commands may change the values associated with a name by previous commands so a new name is not necessarily introduced for each new command. Thus, in order to carry out several commands several times, those commands need not be duplicated. Instead, the same commands are repeated. Hence, each name may be, and usually will be, associated with different values while a program is running. For example, in order to find the sum of the N elements of array A, we do not write:

```
SUM1 := A[1] ;
SUM2 := SUM1 + A[2] ;
SUM3 := SUM2 + A[3] ;
 ...
```

Instead of creating N new SUMs and referring to each element of A explicitly, we write a loop that reuses one name for the sum, say SUM, and another that indicates successive array elements, say I:

```
I := 0 ;
SUM := 0 ;
WHILE I < N DO
BEGIN
        I := I + 1 ;
        SUM := SUM + A[I]
END
```

In imperative languages, new values may be associated with the same name through command repetition.

In functional languages, because the same names cannot be reused with different values, nested function calls are used to create new versions of the names for new values. Similarly, because command repetition cannot be used to change the values associated with names, **recursive** function calls are used repeatedly to create new versions of names associated with new values. Here, a function calls itself to create new versions of its formal parameters which are then bound to new actual parameter values. For example, we might write a function, in a style similar to Pascal, to sum an array:

```
FUNCTION SUM(A:ARRAY [1..N] OF INTEGER;
                    I,N:INTEGER):INTEGER;
BEGIN
      IF I > N THEN
       SUM := 0
      ELSE
       SUM := A[I] + SUM(A,I+1,N)
END
```

Thus, for the function call:

```
SUM(B,1,M)
```

the sum is found through successive recursive calls to SUM:

```
B[1] + SUM(B,2,M) =
B[1] + B[2] + SUM(B,3,M) =
B[1] + B[2] + ... + B[M] + SUM(B,M+1,M) =
B[1] + B[2] + ... + B[M] + 0
```

Here, each recursive call to SUM creates new local versions of A, I and N, and the previous versions become inaccessible. At the end of each recursive call, the new local variables are lost, the partial sum is returned to the previous call and the previous local variables come back into use.

In functional languages, new values are associated with new names through recursive function call nesting.

1.5 Data structures in functional languages

In imperative languages, array elements and record fields are changed by successive assignments. In functional languages, because there is no assignment, substructures in data structures cannot be changed one at a time. Instead, it is necessary to write down a whole structure with explicit changes to the appropriate substructure.

Functional languages provide explicit representations for data structures.

Functional languages do not provide arrays because without assignment there is no easy way to access an arbitrary element. Writing out an entire array with a change to one element would be ludicrously unwieldy. Instead, nested data structures like lists are provided. These are based on recursive notations where operations on a whole structure are described in terms of recursive operations on substructures. The representations for nested data structures are often very similar to the nested function call notation. Indeed, in LISP (LISt Programming) the same representation is used for functions and data structures.

This ability to represent entire data structures has a number of advantages. It provides a standard format for displaying structures which greatly simplifies program debugging and final output as there is no need to write special printing subprograms for each distinct type of structure. It also provides a standard format for storing data structures which can remove the need to write special file I/O subprograms for distinct types of structure.

A related difference lies in the absence of global structures in functional languages. In imperative languages, if a program manipulates single distinct data structures, then it is usual to declare them as globals at the top level of a program. Their substructures may then be accessed and modified directly through assignment within subprograms without passing them as parameters. In functional languages, because there is no assignment, it is not possible to change substructures of global structures independently. Instead, entire data structures are passed explicitly as actual parameters to functions for substructure changes and the entire changed structure is then passed back again to the calling function. This means that function calls in a functional program are larger than their equivalents in an imperative program because of these additional parameters. However, it has the advantage of ensuring that structure manipulation by functions is always explicit in the function definitions and calls. This makes it easier to see the flow of data in programs.

1.6 Functions as values

In many imperative languages, subprograms may be passed as actual parameters to other subprograms but it is rare for an imperative language to allow subprograms to be passed back as results. In functional languages, functions may construct new functions and pass them on to other functions.

Functional languages allow functions to be treated as values.

For example, the following contrived, illegal, Pascal-like function returns an arithmetic function depending on its parameter:

```
TYPE OPTYPE = (ADD,SUB,MULT,QUOT);

FUNCTION ARITH(OP:OPTYPE):FUNCTION;
 FUNCTION SUM(X,Y:INTEGER):INTEGER;
  BEGIN SUM := X+Y END;
 FUNCTION DIFF(X,Y:INTEGER):INTEGER;
  BEGIN DIFF := X-Y END;
 FUNCTION TIMES(X,Y:INTEGER):INTEGER;
  BEGIN TIMES := X*Y END;
 FUNCTION DIVIDE(X,Y:INTEGER):INTEGER;
  BEGIN DIVIDE := X DIV Y END;
BEGIN
      CASE OP OF
         ADD: ARITH := SUM;
         SUB: ARITH := DIFF;
         MULT: ARITH := TIMES;
         QUOT: ARITH := DIVIDE;
      END
END
```

Thus:

```
ARITH(ADD)
```

returns the FUNCTION:

```
SUM
```

and:

```
ARITH(SUB)
```

returns the FUNCTION:

```
DIFF
```

and so on. Thus, we might add two numbers with:

```
ARITH(ADD)(3,4)
```

and so on. This is illegal in many imperative languages because it is not possible to construct functions of type 'function'. As we shall see, the ability to manipulate functions as values gives functional languages substantial power and flexibility.

1.7 The origins of functional languages

Functional programming has its roots in mathematical logic. Informal logical systems have been in use for over 2000 years but the first modern formalizations were by Hamilton, De Morgan and Boole in the mid nineteenth century. Within their works we now distinguish the **propositional calculus** and the **predicate calculus.**

Propositional calculus is a system with true and false as basic values and with and, or, not and so on as basic operations. Names are used to stand for arbitrary truth values. Within propositional calculus, it is possible to **prove** whether or not an arbitrary expression is a **theorem** (always true), by starting with **axioms** (elementary expressions which are always true), and applying **rules of inference** to construct new theorems from axioms and existing theorems. Propositional calculus provides a foundation for more powerful logical systems. It is also used to describe digital electronics where on and off signals are represented as true and false respectively, and electronic circuits are represented as logical expressions.

Predicate calculus extends propositional calculus to enable expressions involving non-logical values like numbers, sets or strings. This is achieved through the introduction of **predicates** which generalize logical expressions to describe properties of non-logical values, and **functions** to generalize operations on non-logical values. It also introduces the idea of **quantifiers** to describe properties of sequences of values, for example,

universal quantification, for all of a sequence having a property, or **existential quantification**, for one of a sequence having a property. Additional axioms and rules of inference are provided for quantified expressions. Predicate calculus may be applied to different problem areas through the development of appropriate predicates, functions, axioms and rules of inference. For example, number theoretic predicate calculus is used to reason about numbers. Functions are provided for arithmetic and predicates are provided for comparing numbers. Predicate calculus also forms the basis of logic programming in languages like PROLOG, and of expert systems based on logical inference.

Note that within propositional and predicate calculi, associations between names and values are unchanging and expressions have no necessary evaluation order.

The late nineteenth century also saw Peano's development of a formal system for **number theory.** This introduced numbers in terms of 0 and the successor function, so any number is that number of successors of 0. Proofs in the system were based on a form of **induction** which is akin to **recursion**.

At the turn of the century, Russell and Whitehead attempted to derive mathematical truth directly from logical truth in their *Principia Mathematica*. They were, in effect, trying to construct a logical description for mathematics. Subsequently, the German mathematician Hilbert proposed a 'Program' to demonstrate that *Principia* really did describe mathematics totally. He required proof that the *Principia* description of mathematics was *consistent,* that is, it did not allow any contradictions to be proved, and *complete,* that is , it allowed every true statement of number theory to be proved. Alas, in 1931, Godel showed that any system powerful enough to describe arithmetic was necessarily incomplete.

However, Hilbert's 'Program' had promoted vigorous investigation into the **theory of computability,** to try to develop formal systems to describe computations in general. In 1936, three distinct formal approaches to computability were proposed: Turing's **Turing machines**, Kleene's **recursive function theory** (based on Hilbert's work from 1925) and Church's **λ calculus.** Each is well defined in terms of a simple set of primitive operations and a simple set of rules for structuring operations; most important, each has a proof theory.

All the above approaches have been shown formally to be equivalent to each other and also to generalized von Neumann machines – digital computers. This implies that a result from one system will have equivalent results in equivalent systems and that any system may be used to model any other system. In particular, any results will apply to digital computer languages and any of these systems may be used to describe computer languages. Contrariwise, computer languages may be used to describe and hence implement any of these systems. Church hypothesized that all descriptions of computability are equivalent. While **Church's thesis** cannot be proved formally, every subsequent description of computability has been proved to be equivalent to existing descriptions.

An important difference between Turing machines, recursive functions and λ calculus is that the Turing machine approach concentrated on computation as mechanized symbol manipulation based on assignment and time ordered evaluation, whereas recursive function theory and λ calculus emphasized computation as structured function application. They both are evaluation order independent.

Turing demonstrated that it is impossible to tell whether or not an arbitrary Turing machine will halt – the **halting problem** is unsolvable. This also applies to λ calculus and recursive function theory, so there is no way of telling if evaluation of an arbitrary λ expression or recursive function call will ever terminate. However, Church and Rosser showed for λ calculus that if different evaluation orders do terminate then the results will be the same. They also showed that one particular evaluation order is more likely to lead to termination than any other. This has important implications for computing because it may be more efficient to carry out some parts of a program in one order and other parts in another order. In particular, if a language is evaluation order independent, then it may be possible to carry out program parts in parallel.

Today, λ calculus and recursive function theory are the backbones of functional programming but they have wider applications throughout computing.

1.8 Computing and the theory of computing

The development of electronic digital computers in the 1940s and 1950s led to the introduction of high level languages to simplify programming. Computability theory had a direct impact on programming language design. For example, Algol 60, an early general purpose high level language and an ancestor of Pascal, had recursion and the λ calculus based call-by-name parameter passing mechanism.

As computer use increased dramatically in the 1960s, there was renewed interest in the application of formal ideas about computability to practical computing. In 1963, McCarthy proposed a mathematical basis for computation which was influenced by λ calculus and recursive function theory. This culminated in the LISP programming language. LISP is a very simple language based on recursive functions manipulating lists of words and numbers. Variables are not typed so there are no restrictions on which values may be associated with names. There is no necessary distinction between programs and data – a LISP program is a list. This makes it easy for programs to manipulate programs. LISP was one of the first programming languages with a rigorous definition but it is not a pure functional language and contains imperative as well as functional elements. It has had a lot of influence on functional language design and functional programming techniques. At present, LISP is used mainly within the artificial intelligence community but there is growing industrial interest in it. McCarthy also introduced techniques for proving recursive function based programs.

In the mid 1960s, Landin and Strachey both proposed the use of λ calculus to model imperative languages. Landin's approach was based on an **operational** description of λ calculus defined in terms of an **abstract interpreter** for it – the SECD machine. Having described λ calculus, Landin then used it to construct an abstract interpreter for Algol 60. (McCarthy had also used an abstract interpreter to describe LISP.) This approach formed the basis of the Vienna Definition Language (VDL) which was used to define IBM's PL/I. The SECD machine has been adapted to implement many functional languages on digital computers. Landin also developed the pure functional language ISWIM which influenced later languages. Strachey's approach was to construct descriptions of imperative languages using a notation based on λ calculus so that every imperative language construct would have an equivalent function **denotation.** This approach was strengthened by Scott's **lattice theoretic** description for λ calculus. Currently, **denotational semantics** and its derivatives are used to give formal definitions of programming languages. Functional languages are closely related to λ calculus based semantic languages.

Since the introduction of LISP, many partially and fully functional languages have been developed. For example, POP-2 was developed at Edinburgh University by Popplestone and Burstall in 1971 as an updated LISP with a modern syntax and a pure functional subset. It has led to POP11 and to POPLOG which combines POP11 and PROLOG. SASL, developed by Turner at St Andrews University in 1974, is based strongly on λ calculus. Its offspring include KRC and Miranda. Miranda is used as a general purpose language in research and teaching. Hope was developed by Burstall at Edinburgh University in 1980 and is used as the programming language for the ALICE parallel computer. ML was developed by Milner at Edinburgh University in 1979 as a language for the computer assisted proof system LCF. Standard ML is now used as a general purpose functional language. Like LISP, it has imperative extensions.

Interest in functional languages was increased by a paper by Backus in 1978. He claimed that computing was restricted by the structure of digital computers and imperative languages, and proposed the use of Functional Programming (FP) systems for program development. FP systems are very simple, consisting of basic atomic objects and operations, and rules for structuring them. They depend strongly on the use of functions which manipulate other functions as values. They have solid theoretical foundations and are well suited to program proof and refinement. They also have all the time order independence properties that we considered previously. FP systems have somewhat tortuous syntax and are not as easy to use as other functional languages. However, Backus' paper has been very influential in motivating the broader use of functional languages.

In addition to the development of functional languages, there is considerable research into formal descriptions of programming languages using techniques related to λ calculus and recursive function theory. This is both theoretical, to develop and extend formalisms and proof systems, and practical, to form the basis of programming methodologies and language implementations. Major areas where computing theory has practical applications include the precise specification of programs, the development of prototypes from specifications and the ability to verify that implementations correspond to specifications. For example, the Vienna Development Method (VDM), Z and OBJ approaches to program specification all use functional notations, and functional language implementations are used for prototyping. Proof techniques related to recursive function theory, for example constructive type theory, are used to verify programs and to construct correct programs from specifications.

1.9 λ calculus

The λ calculus is a surprisingly simple yet powerful system. It is based on function **abstraction,** to generalize expressions through the introduction of names, and function **application,** to evaluate generalized expressions by giving names particular values.

The λ calculus has a number of properties which suit it well for describing programming languages. Firstly, only abstraction and application are needed to develop representations for arbitrary programming language constructs; thus λ calculus can be treated as a universal machine code for programming languages. In particular, because λ calculus is evaluation order independent, it can be used to describe and investigate the implications of different evaluation orders in different programming languages. Secondly, there are well developed proof techniques for λ calculus and these can be applied to λ calculus language descriptions of other languages. Finally, because λ calculus is very simple, it is relatively easy to implement. Thus a λ calculus description of language can be used as a prototype and run on a λ calculus implementation to try the language out. We will not consider proof or implementation further here.

In this book we are going to use λ calculus to explore functional programming. Pure λ calculus does not look much like a programming language. Indeed, all it provides are names, function abstraction and function application. However, it is straightforward to develop new language constructs from this basis. Here we will use λ calculus to construct step-by-step a compact, general purpose functional programming notation.

SUMMARY

- Imperative languages are based on assignment sequences whereas functional languages are based on nested function calls.
- In imperative languages, the same name may be associated with several values, whereas in functional languages a name is only associated with one value.
- Imperative languages have fixed evaluation orders whereas functional languages need not.
- In imperative languages, new values may be associated with the same name through command repetition whereas in functional languages new names are associated with new values through recursive function call nesting.
- Functional languages provide explicit data structure representations.
- In functional languages, functions are values.
- Functional languages originate in mathematical logic and the theory of computing, in recursive function theory and λ calculus.

Chapter 2
λ calculus

In this chapter we are going to meet λ calculus, which will be used as a basis for functional programming in the rest of the book. To begin with, we will briefly discuss abstraction as generalization through the introduction of names in expressions, and specialization through the replacement of names with values. The role of abstraction in programming languages will also be considered. Next, we will take an overview of λ calculus as a system for abstractions based on functions and function applications. The rules for constructing and evaluating λ expressions will be discussed informally in some detail. A notation for defining named functions will be introduced, and we will look at how functions may be constructed from other functions. We will also consider functions for manipulating pairs of values. These will be used as building blocks in subsequent chapters to add syntax for booleans, numbers, types and lists to λ calculus. Finally, we will take a slightly more formal look at λ expression evaluation.

2.1 Abstraction

Abstraction is central to problem solving and programming. It involves generalization from concrete instances of a problem so that a general solution may be formulated. A general, abstract solution may then be used in turn to solve particular, concrete instances of the problem.

The simplest way to specify an instance of a problem is in terms of particular concrete operations on particular concrete objects. Abstraction is based on the use of names to stand for concrete objects and operations, to generalize the instances. A generalized instance may subsequently be turned into a particular instance by replacing the names with new concrete objects and operations.

We will try to get a feel for abstraction through a somewhat contrived example. Consider buying 9 items at 10 cents each. The total cost is:

```
10*9
```

Here, we are carrying out the concrete operation of multiplication on the concrete values 10 and 9. Now consider buying 11 items at 10 cents each. The total cost is:

```
10*11
```

Here we are carrying out the concrete operation of multiplication on the concrete values 10 and 11. We can see that as the number of items changes so the formula for the total cost changes at the place where the number of items appears. We can **abstract over** the number of items in the formula by introducing a name to stand for a general number of items, say items:

```
10*items
```

We might make this abstraction explicit by preceding the formula with the name used for abstraction:

```
REPLACE items IN 10*items
```

Here, we have abstracted over an operand in the formula. To evaluate this abstraction, we need to supply a value for the name. For example, for 84 items:

```
REPLACE items WITH 84 IN 10*items
```

which gives:

```
10*84
```

We have made a **function** from a formula by replacing an object with a name and identifying the name that we used. We have then **evaluated** the function by replacing the name in the formula with a new object and evaluating the resulting formula.

Let us use abstraction again to generalize our example further. Suppose the cost of items goes up to 11 cents. Now the total cost of items is:

```
REPLACE items IN 11*items
```

Suppose the cost of items drops to 9 cents. Now the total cost of items is:

```
REPLACE items IN 9*items
```

Because the cost changes, we could also introduce a name to stand for the cost in general, say cost:

```
REPLACE cost IN
 REPLACE items IN cost*items
```

Here, we have abstracted over two operands in the formula. To evaluate the abstraction, we need to supply two values. For example, 12 items at 32 cents will have total cost:

```
REPLACE cost WITH 32 IN
 REPLACE items WITH 12 IN cost*items
```

which is:

```
REPLACE items WITH 12 IN 32*items
```

which is:

```
32*12
```

For example, 25 items at 15 cents will have total cost:

```
REPLACE cost WITH 15 IN
 REPLACE items WITH 25 IN cost*items
```

which is:

```
REPLACE items WITH 25 IN 15*items
```

which is:

```
15*25
```

Suppose we now want to solve a different problem. We are given the total cost and the number of items and we want to find out how much each item costs. For example, if 12 items cost 144 cents, then each item costs:

```
144/12
```

If 15 items cost 45 cents then each item costs:

```
45/15
```

In general, if items items cost cost cents then each item costs:

```
REPLACE cost IN
 REPLACE items IN cost/items
```

Now, compare this with the formula for finding a total cost:

```
REPLACE cost IN
 REPLACE items IN cost*items
```

They are the same except for the operation '/' in finding the cost of each item and '*' in finding the cost of all items. We have two instances of a problem involving the application of an operation to two operands.

We could generalize these instances by introducing a name, say op, where the operation is used:

```
REPLACE op IN
 REPLACE cost IN
  REPLACE items IN cost op items
```

Finding the total cost will now require the replacement of the operation name with the concrete multiplication operation:

```
REPLACE op WITH * IN
 REPLACE cost IN
  REPLACE items IN cost op items
```

which is:

```
REPLACE cost IN
 REPLACE items IN cost * items
```

Similarly, finding the cost of each item will require the replacement of the operation name with the concrete division operation:

```
REPLACE op WITH / IN
 REPLACE cost  IN
  REPLACE item IN cost op items
```

which is:

```
REPLACE cost IN
 REPLACE items IN cost / items
```

Abstraction is based on generalization through the introduction of a name to replace a value and specialization through the replacement of a name with another value.

Note that care must be taken with generalization and specialization to ensure that names are replaced by objects and operations of appropriate types. In the above examples, the operand names must be replaced by numbers and the operator name must be replaced by an operation with two number arguments. We will look at this in slightly more detail in Chapter 5.

2.2 Abstraction in programming languages

Abstraction lies at the heart of all programming languages. In imperative languages, variables as name/value associations are abstractions for computer memory locations based on specific address/value associations. The particular address for a variable is irrelevant so long as the name/value association is consistent. Indeed, on a computer with memory management, a variable will correspond to many different concrete locations as a program's data space is swapped in and out of different physical memory areas. The compiler and the run time system make sure that variables are implemented as consistent, concrete locations.

Where there are abstractions, there are mechanisms for both introducing and specializing them. For example, in Pascal, variables are introduced with declarations and given values by statements for use in subsequent statements. Variables are then used as abstractions for memory addresses on the left of assignment statements or in READ statements, and as abstractions for values in expressions on the right of assignment statements or in WRITE statements.

Abstractions may be subject to further abstraction. This is the basis of hierarchical program design methodologies and modularity. For example, Pascal procedures are abstractions for sequences of statements, named by procedure declarations, and functions are abstractions for expressions, named by function declarations. Procedures and functions declare formal parameters which identify the names used to abstract in statement sequences and expressions. Simple, array and record variable formal parameters are abstractions for simple, array and record variables in statements and expressions. Procedure and function formal parameters are abstractions for procedures and functions in statements and expressions. Actual parameters specialize procedures and functions. Procedure calls with actual parameters invoke sequences of statements with formal parameters replaced by actual parameters. Similarly, function calls with actual parameters evaluate expressions with formal parameters replaced by actual parameters.

Programming languages may be characterized and compared in terms of the abstraction mechanisms they provide. Consider, for example, Pascal and BASIC. Pascal has distinct INTEGER and REAL variables as abstractions for integer and real numbers, whereas BASIC has only numeric variables which abstract over both. Pascal also has CHAR variables as abstractions for single letters. Both Pascal and BASIC have arrays which are abstractions for sequences of variables of the same type. BASIC has string variables as abstractions for letter sequences, whereas in Pascal an array of CHARs is used. Pascal also has records as abstractions for sequences of variables of differing types. Pascal has procedures and functions as statement and expression abstractions. Furthermore, procedure and function formal parameters abstract over procedures and functions within procedures and functions. The original BASIC only had functions as expression abstractions and did not allow function formal parameters to abstract over functions in expressions. In subsequent chapters, we will see how abstraction may be used to define many aspects of programming languages.

2.3 Introducing λ calculus

The λ calculus was devised by Alonzo Church in the 1930s as a model for computability and has subsequently been central to contemporary computer science. It is a very simple but very powerful language based on pure abstraction. It can be used to formalize all aspects of programming languages and programming, and is particularly suited for use as a 'machine code' for functional languages and functional programming.

In the rest of this chapter we are going to look at how λ calculus expressions are written and manipulated. This may seem a bit disjointed at first: it is hard to introduce all of a new topic simultaneously and so some details will be rather sketchy to begin with. A set of useful functions will be built up bit by bit. The functions introduced in this chapter to illustrate various aspects of λ calculus will be used as building blocks in later chapters. Each example may assume knowledge of previous examples and so it is important that you work through the material slowly and consistently.

2.4 λ expressions

The λ calculus is a system for manipulating λ **expressions.** A λ expression may be a **name** to identify an abstraction point, a **function** to introduce an abstraction or a **function application** to specialize an abstraction:

```
<expression> ::= <name> | <function> | <application>
```

A name may be any sequence of non-blank characters, for example:

```
fred   legs-11   19th_nervous_breakdown   33   +   -->
```

A λ function is an abstraction over a λ expression and has the form:

```
<function> ::= λ<name>.<body>
```

where:

```
<body> ::= <expression>
```

for example:

```
λx.x   λfirst.λsecond.first   λf.λa.(f a)
```

The λ precedes and introduces a name used for abstraction. The name is called the function's **bound variable** and is like a formal parameter in a Pascal function declaration. The '.' separates the name from the expression in which abstraction with that name takes place. This expression is called the function's **body.** Notice that the body expression may be any λ expression including another function. This is far more general than, for example, Pascal, which does not allow functions to return functions as values. Note that functions do not have names! For example, in Pascal, the function name is always used to refer to the function's definition.

A function application has the form:

```
<application> ::= (<function expression> <argument expression>)
```

where:

```
<function expression> ::= <expression>
<argument expression> ::= <expression>
```

for example:

```
(λx.x λa.λb.b)
```

A function application specializes an abstraction by providing a value for the name. The function expression contains the abstraction to be specialized with the argument expression.

In a function application, also known as a **bound pair**, the function expression is said to be **applied to** the argument expression. This is like a function call in Pascal where the argument expression corresponds to the actual parameter. The crucial difference is that in Pascal the function name is used in the function call and the implementation picks up the corresponding definition; the λ calculus is far more general and allows function definitions to appear directly in function calls.

There are two approaches to evaluating function applications. For both, the function expression is evaluated to return a function. All occurrences of the function's bound variable in the function's body expression are then replaced by *either*

the value of the argument expression

or

the unevaluated argument expression.

Finally, the function body expression is evaluated.

The first approach is called **applicative order** reduction and is like Pascal 'call-by-value': the actual parameter expression is evaluated before being passed to the formal parameter. The second approach is called **normal order** reduction and is like 'call-by-name' in Algol 60: the actual parameter expression is not evaluated before being passed to the formal parameter. As we will see, normal order is more powerful than applicative order but may be less efficient. For now, all function applications will be evaluated in normal order.

The syntax allows a single name as a λ expression but in general we will restrict single names to the bodies of functions. This is so that we can avoid having to consider names as objects in their own right, for example LISP or PROLOG literals, as it complicates the presentation. We will discuss this further later on.

2.5 Simple λ functions

We will now look at a variety of simple λ functions.

2.5.1 Identity function

Consider the function:

    λx.x

This is the identity function which returns the argument to which it is applied. Its bound variable is:

    x

and its body expression is the name:

    x

When it is used as the function expression in a function application, the bound variable x will be replaced by the argument expression in the body expression x, giving the original argument expression. Suppose the identity function is applied to itself:

    (λx.x λx.x)

This is a function application with:

    λx.x

as the function expression and:

    λx.x

as the argument expression. When this application is evaluated, the bound variable:

    x

for the function expression:

    λx.x

is replaced by the argument expression:

```
λx.x
```

in the body expression:

```
x
```

giving:

```
λx.x
```

An identity operation always leaves its argument unchanged. In arithmetic, adding or subtracting 0 are identity operations. For any number <number>:

```
<number> + 0 = <number>
<number> - 0 = <number>
```

Multiplying or dividing by 1 are also identity operations:

```
<number> * 1 = <number>
<number> / 1 = <number>
```

The identity function is an identity operation for λ functions.

We could equally well have used different names for the bound variable, for example:

```
λa.a
```

or:

```
λyibble.yibble
```

to define other versions of the identity function. We will consider naming in more detail later but note just now that we can consistently change names.

2.5.2 Self-application function

Consider the rather odd function:

```
λs.(s s)
```

which applies its argument to its argument. The bound variable is:

s

and the body expression is the function application:

(s s)

which has the name:

s

as function expression and the same name:

s

as argument expression.

Let us apply the identity function to it:

(λx.x λs.(s s))

In this application, the function expression is:

λx.x

and the argument expression is:

λs.(s s)

When this application is evaluated, the function expression bound variable:

x

is replaced by the argument:

λs.(s s)

in the function expression body:

x

giving:

λs.(s s)

which is the original argument.

Let us apply this self-application function to the identity function:

```
(λs.(s s) λx.x)
```

Here, the function expression is:

```
λs.(s s)
```

and the argument expression is:

```
λx.x
```

When this application is evaluated, the function expression bound variable:

```
s
```

is replaced by the argument:

```
λx.x
```

in the function expression body:

```
(s s)
```

giving a new application:

```
(λx.x λx.x)
```

with function expression:

```
λx.x
```

and argument expression:

```
λx.x
```

This is now evaluated as above giving the final value:

```
λx.x
```

Consider the application of the self-application function to itself:

```
(λs.(s s) λs.(s s))
```

This application has function expression:

    λs.(s s)

and argument expression:

    λs.(s s)

To evaluate it, the function expression bound variable:

    s

is replaced by the argument:

    λs.(s s)

in the function expression body:

    (s s)

giving a new application:

    (λs.(s s) λs.(s s))

with function expression:

    λs.(s s)

and argument expression:

    λs.(s s)

which is then evaluated. The function expression bound variable:

    s

is replaced by the argument:

    λs.(s s)

in the function expression body:

    (s s)

giving the new application:

```
(λs.(s s) λs.(s s))
```

which is then evaluated...

Each application evaluates to the original application so this application never terminates! A version of this self-application function is used to construct recursive functions in Chapter 4. Here, it should be noted that not all expression evaluations terminate. In fact, as we will note in Chapter 8, there is no way of telling whether or not an expression evaluation will ever terminate!

2.5.3 Function application function

Consider the function:

```
λfunc.λarg.(func arg)
```

This has bound variable:

```
func
```

and the body expression is another function:

```
λarg.(func arg)
```

which has bound variable:

```
arg
```

and a function application:

```
(func arg)
```

as body expression. This in turn has the name:

```
func
```

as function expression and the name:

```
arg
```

as argument expression. When used, the whole function returns a second function which then applies the first function's argument to the second function's argument. For example, let us use it to apply the identity function to the self-application function:

```
((λfunc.λarg.(func arg) λx.x) λs.(s s))
```

In this application, the function expression is itself an application:

```
(λfunc.λarg.(func arg) λx.x)
```

which must be evaluated first. The bound variable:

```
func
```

is replaced by the argument:

```
λx.x
```

in the body:

```
λarg.(func arg)
```

giving:

```
λarg.(λx.x arg)
```

which is a new function which applies the identity function to its argument. The original expression is now:

```
(λarg.(λx.x arg) λs.(s s))
```

and so the bound variable:

```
arg
```

is replaced by the argument:

```
λs.(s s)
```

in the body:

```
(λx.x arg)
```

giving:

```
(λx.x λs.(s s))
```

which is now evaluated as above. The bound variable:

```
x
```

is replaced by the argument:

```
λs.(s s)
```

in the body:

```
x
```

giving:

```
λs.(s s)
```

2.6 Introducing new syntax

As our λ expressions become more elaborate, they become harder to manipulate. To simplify working with λ expressions and to construct a higher level functional language, we will allow the use of more concise notations. For example, in this and subsequent chapters we will introduce named function definitions, infix operations, an IF style conditional expression and so on. This addition of higher level layers to a language is known as **syntactic sugaring** because the representation of the language is changed but the underlying meaning stays the same.

New syntax will be introduced for commonly used constructs through **substitution rules**. The application of these rules will not involve making choices. Their use will lead to pure λ expressions after a finite number of steps involving simple substitutions. This is to ensure that we can always 'compile' completely a higher level representation into λ calculus before evaluation. Then, we only need to refer to our original simple λ calculus rules for evaluation. In this way we will not need to modify or augment the λ calculus itself and, should we need to, we can rely on the existing theories for λ calculus without developing them further.

Furthermore, we are going to use λ calculus as a time order independent language to investigate time ordering. Thus, our substitution rules should also be time order independent. Otherwise, different substitution orders might lead to different substitutions being made and these might result in expressions with different meanings. The simplest way to ensure time order independence is to insist that all substitutions be made statically or be capable of being made statically. We can then apply them to produce pure λ calculus before evaluation starts.

We will not always make all substitutions before evaluation as this would lead to pages and pages of incomprehensible λ expressions for large higher level expressions. *We will insist, however, that making all substitutions before evaluation always remains a possibility.*

2.7 Notations for naming functions and β reduction

It is a bit tedious writing functions out repeatedly. We will now name functions using:

```
def <name> = <function>
```

to define a name/function association. For example, we could name the functions we looked at in the previous sections:

```
def identity = λx.x
def self_apply = λs.(s s)
def apply = λfunc.λarg.(func arg)
```

Now we can use the <name> in expressions to stand for the <function>.

Strictly speaking, all defined names in an expression should be replaced by their definitions before the expression is evaluated. However, for now we will only replace a name by its associated function when the name is the function expression of an application. We will use the notation:

```
(<name> <argument>) == (<function> <argument>)
```

to indicate the replacement of a <name> by its associated <function>.

Formally, the replacement of a bound variable with an argument in a function body is called **beta reduction** (**β reduction**). In future, instead of detailing each normal order β reduction we will introduce the notation:

```
(<function> <argument>) => <expression>
```

to mean that the <expression> results from the application of the <function> to the unevaluated <argument>. When we have seen a sequence of reductions before, or we are familiar with the functions involved, we will omit the reductions and write:

```
=> ... =>
```

to show where they should be.

2.8 Functions from functions

We can use the self-application function to build versions of other functions. For example, let us define a function with the same effect as the identity function:

```
def identity2 = λx.((apply identity) x)
```

Let us apply this to the identity function:

```
(identity2 identity) ==
(λx.((apply identity) x) identity) =>
((apply identity) identity) ==
((λfunc.λarg.(func arg) identity) identity) =>
(λarg.(identity arg) identity) =>
(identity identity) => ... =>
identity
```

Let us show that identity and identity2 are equivalent. Suppose:

```
<argument>
```

stands for any expression. Then:

```
(identity2 <argument>) ==
(λx.((apply identity) x) <argument>) =>
((apply identity) <argument>) => ... =>
(identity <argument>) => ... =>
<argument>
```

so identity and identity2 have the same effect.

We can use the function application function to define a function with the same effect as the function application function itself. Suppose:

```
<function>
```

is any function. Then:

```
(apply <function>) ==
(λfunc.λarg.(func arg) <function>) =>
λarg.(<function> arg)
```

Applying this to any argument:

```
<argument>
```

we get:

```
(λarg.(<function> arg) <argument>) =>
(<function> <argument>)
```

which is the application of the original function to the argument. Using apply adds a layer of β reduction to an application.

We can also use the function application function slightly differently to define a function with the same effect as the self-application function:

```
def self_apply2 = λs.((apply s) s)
```

Let us apply this to the identity function:

```
(self_apply2 identity) ==
(λs.((apply s) s) identity) =>
((apply identity) identity) => ... =>
(identity identity) => ... =>
identity
```

In general, applying self_apply2 to any argument:

```
<argument>
```

gives:

```
(self_apply2 <argument>) ==
(λs.((apply s) s) <argument>) =>
((apply <argument>) <argument>) => ... =>
(<argument> <argument>)
```

so self_apply and self_apply2 have the same effect.

2.9 Argument selection and argument pairing functions

We are now going to look at functions for selecting arguments in nested function applications. We will use these functions a great deal in later stages to model boolean logic, integer arithmetic and list data structures.

2.9.1 Selecting the first of two arguments

Consider the function:

```
def select_first = λfirst.λsecond.first
```

This function has bound variable:

```
first
```

and body:

```
λsecond.first
```

When applied to an argument, it returns a new function which, when applied to another argument, returns the first argument. For example:

```
((select_first identity) apply) ==
((λfirst.λsecond.first identity) apply) =>
(λsecond.identity apply) =>
identity
```

In general, applying select_first to arbitrary arguments:

```
<argument1>
```

and:

```
<argument2>
```

returns the first argument:

```
((select_first <argument1>) <argument2>) ==
((λfirst.λsecond.first <argument1>) <argument2>) =>
(λsecond.<argument1> <argument2>) =>
<argument1>
```

Note that select_first's body contains no references to second, so the second argument is never used.

2.9.2 Selecting the second of two arguments

Consider the function:

```
def select_second = λfirst.λsecond.second
```

This function has bound variable:

first

and body:

λsecond.second

which is another version of the identity function. When applied to an argument select_second returns a new function which, when applied to another argument, returns the other argument. For example:

((select_second identity) apply) ==

((λfirst.λsecond.second identity) apply) =>

(λsecond.second apply) =>

apply

The first argument identity was lost because the bound variable first does not appear in the body λsecond.second.

In general, applying select_second to arbitrary arguments:

<argument1>

and:

<argument2>

returns the second argument:

((select_second <argument1>) <argument2>) ==

((λfirst.λsecond.second <argument1>) <argument2>) =>

((second.second <argument2>) =>

<argument2>

We can show that select_second applied to anything returns a version of identity. As before, we will use:

<argument>

to stand for an arbitrary expression, so:

(select_second <argument>) ==

(λfirst.λsecond.second <argument>) =>

λsecond.second

If second is replaced by x then:

```
λsecond.second
```

becomes:

```
λx.x
```

Notice that select_first applied to identity returns a version of select_second:

```
(select_first identity) ==
(λfirst.λsecond.first identity) =>
λsecond.identity ==
λsecond.λx.x
```

If second is replaced by first and x by second then this becomes:

```
λfirst.λsecond.second
```

2.9.3 Making pairs from two arguments

Consider the function:

```
def make_pair = λfirst.λsecond.λfunc.((func first) second)
```

with bound variable:

```
first
```

and body:

```
λsecond.λfunc.((func first) second)
```

This function applies argument func to argument first to build a new function which may be applied to argument second. Note that arguments first and second are used before argument func to build a function:

```
λfunc.((func first) second)
```

Now, if this function is applied to select_first then argument first is returned and if it is applied to select_second then argument second is returned. For example:

```
((make_pair identity) apply) ==
((λfirst.λsecond.λfunc.((func first) second) identity) apply) =>
(λsecond.λfunc.((func identity) second) apply) =>
λfunc.((func identity) apply)
```

Now, if this function is applied to select_first:

```
(λfunc.((func identity) apply) select_first) ==
((select_first identity) apply) ==
((λfirst.λsecond.first identity) apply) =>
(λsecond.identity apply) =>
identity
```

and if it is applied to select_second:

```
(λfunc.((func identity) apply) select_second) ==
((select_second identity) apply) ==
((λfirst.λsecond.second identity) apply) =>
(λsecond.second apply) =>
apply
```

In general, applying make_pair to arbitrary arguments:

```
<argument1>
```

and:

```
<argument2>
```

gives:

```
((make_pair <argument1>) <argument2>) ==
((λfirst.λsecond.λfunc.((func first) second) <argument1>)
  <argument2>) =>
(λsecond.λfunc.((func <argument1>) second) <argument2>) =>
λfunc.((func <argument1>) <argument2>)
```

Thereafter, applying this function to select_first returns the first argument:

```
(λfunc.((func <argument1>) <argument2>) select_first) =>
((select_first <argument1>) <argument2>) ==
((λfirst.λsecond.first <argument1>) <argument2>) =>
(λsecond.<argument1> <argument2>) =>
<argument1>
```

and applying this function to select_second returns the second argument:

```
(λfunc.((func <argument1>) <argument2>) select_second) =>
((select_second <argument1>) <argument2>) ==
((λfirst.λsecond.second <argument1>) <argument2>) =>
(λsecond.second <argument2>) =>
<argument2>
```

2.10 Free and bound variables

We are now going to consider how we ensure that arguments are substituted correctly for bound variables in function bodies. If all the bound variables for functions in an expression have distinct names, then there is no problem. For example, in:

```
(λf.(f λx.x) λs.(s s))
```

there are three functions. The first has bound variable f, the second has bound variable x and the third has bound variable s. Thus:

```
(λf.(f λx.x) λs.(s s)) =>
(λs.(s s) λx.x) =>
(λx.x λx.x) =>
λx.x
```

It is possible, however, for bound variables in different functions to have the same name. Consider:

```
(λf.(f λf.f) λs.(s s))
```

This should give the same result as the previous expression. Here, the bound variable f should be replaced by:

```
λs.(s s)
```

Note that we should replace the first f in:

```
(f λf.f)
```

but not the f in the body of:

```
λf.f
```

This is a new function with a new bound variable which just happens to have the same name as a previous bound variable. To clarify this we need to be more specific about how bound variables relate to variables in function bodies. For an arbitrary function:

```
λ<name>.<body>
```

the bound variable <name> may correspond to occurrences of <name> in <body> and nowhere else. Formally, the **scope** of the bound variable <name> is <body>. For example, in:

```
λf.λs.(f (s s))
```

the bound variable f is in scope in:

```
λs.(f (s s))
```

In:

```
(λf.λg.λa.(f (g a)) λg.(g g))
```

the leftmost bound variable f is in scope in:

```
λg.λa.(f (g a))
```

and nowhere else. Similarly, the rightmost bound variable g is in scope in:

```
(g g)
```

and nowhere else. Note that we have said *may correspond*. This is because the re-use of a name may alter a bound variable's scope, as we will see.

Now we can introduce the idea of a variable being **bound** or **free** in an expression. A variable is said to be bound to occurrences in the body of a function for which it is the bound variable, provided no other functions within the body introduce the same bound variable; otherwise it is said to be free. Thus, in the expression:

```
λx.x
```

the variable x is bound, but in the expression:

```
x
```

the variable x is free. In:

```
λf.(f λx.x)
```

the variable f is bound, but in the expression:

```
(f λx.x)
```

the variable f is free.

In general, for a function:

```
λ<name>.<body>
```

<name> refers to the same variable throughout <body> except where another function has <name> as its bound variable. References to <name> in the new function's body then correspond to the new bound variable and not the old.

In formal terms, all the free occurrences of <name> in <body> are references to the same bound variable <name> introduced by the original function. <name> is in scope in <body> wherever it may occur free, except where another function introduces it in a new scope. For example, in the body of:

```
λf.(f λf.f)
```

which is:

```
(f λf.f)
```

the first f is free so it corresponds to the original bound variable f but subsequent fs are bound and so are distinct from the original bound variable. The outer f is in scope except in the scope of the inner f.

In the body of:

```
λg.((g λh.(h (g λh.(h λg.(h g))))) g)
```

which is:

```
(g λh.(h (g λh.(h λg.(h g))))) g)
```

the first, second and last occurrences of g occur free so they correspond to the outer bound variable g. The third and fourth gs are bound and so are distinct from the original g. The outer g is in scope in the body except in the scope of the inner g.

Let us tighten up our definitions. A variable is **bound** in an expression if:

(1) The expression is an application:

```
(<function> <argument>)
```

and the variable is bound in <function> or <argument>. For example, convict is bound in:

```
(λconvict.convict fugitive)
```

and in:

```
(λprison.prison λconvict.convict)
```

(2) The expression is a function:

```
λ<name>.<body>
```

and either the variable's name is <name> or it is bound in <body>. For example, prisoner is bound in:

```
λprisoner.(number6 prisoner)
```

and in:

```
λprison.λprisoner.(prison prisoner)
```

Similarly, a variable is **free** in an expression if:

(1) The expression is a single name:

```
<name>
```

and the variable's name is <name>. For example, truant is free in:

```
truant
```

(2) The expression is an application:

(<function> <argument>)

and the variable is free in <function> or in <argument>. For example, escaper is free in:

(λprisoner.prisoner escaper)

and in:

(escaper λjailor.jailor)

(3) The expression is a function:

λ<name>.<body>

and the variable's name is not <name> and the variable is free in <body>. For example, fugitive is free in:

λprison.(prison fugitive)

and in:

λshort.λsharp.λshock.fugitive

Note that a variable may be bound and free in different places in the same expression.

We can now define normal order β reduction more formally. In general, for the normal order β reduction of an application:

(λ<name>.<body> <argument>)

we replace all free occurrences of <name> in <body> with <argument>. This ensures that only those occurrences of <name> which actually correspond to the bound variable are replaced. For example, in:

(λf.(f λf.f) λs.(s s))

the first occurrence of f in the body:

(f λf.f)

is free so it is replaced:

(λs.(s s) λf.f) =>

(λf.f λf.f) =>

λf.f

In subsequent examples we will use distinct bound variables.

2.11 Name clashes and α conversion

We have restricted the use of names in expressions to the bodies of functions. This may be restated as the requirement that there be no free variables in a λ expression. Without this restriction, names become objects in their own right. This eases data representation: atomic objects may be represented directly as names, and structured sequences of objects as nested applications using names. However, it also makes reduction much more complicated. For example, consider the function application function:

```
def apply = λfunc.λarg.(func arg)
```

Consider:

```
((apply arg) boing) ==
((λfunc.λarg.(func arg) arg) boing)
```

Here, arg is used both as a function bound variable name and as a free variable name in the leftmost application. These are two distinct uses: the bound variable will be replaced by β reduction but the free variable stays the same. However, if we carry out β reduction literally:

```
((λfunc.λarg.(func arg) arg) boing) =>
(λarg.(arg arg) boing) =>
(boing boing)
```

which was not intended at all. The argument arg has been substituted in the scope of the bound variable arg and appears to create a new occurrence of that bound variable. We can avoid this using consistent renaming. Here, we might replace the bound variable arg in the function with, say arg1:

```
((λfunc.λarg1.(func arg1) arg) boing) =>
(λarg1.(arg arg1) boing) =>
(arg boing)
```

A name clash arises when a β reduction places an expression with a free variable in the scope of a bound variable with the same name as the free variable. Consistent renaming, which is known as **alpha conversion** (α **conversion**), removes the name clash. For a function:

λ<name1>.<body>

the name <name1> and all free occurrences of <name1> in <body> may be replaced by a new name <name2> provided <name2> is not the name of a free variable in λ<name1>.<body>. Note that replacement includes the name at:

λ<name1>

In subsequent examples we will avoid name clashes.

2.12 Simplification through η reduction

Consider an expression of the form:

λ<name>.(<expression> <name>)

This is similar to the function application function above, after application to a function expression only. This is equivalent to:

<expression>

because the application of this expression to an arbitrary argument:

<argument>

gives:

(λ<name>.(<expression> <name>) <argument>) =>

(<expression> <argument>)

This simplification of:

λ<name>.(<expression> <name>)

to:

<expression>

is called **eta reduction** (**η reduction**). We will use it in later chapters to simplify expressions.

SUMMARY

- Abstraction has a central role in programming languages, and λ calculus is a language based on pure abstraction.
- λ calculus syntax has been introduced and used to analyse the structure of some simple expressions.
- Normal order β reduction is used to reduce simple expressions. Note that not all reductions terminate.
- Notations have been introduced for defining functions and simplifying familiar reduction sequences.
- Functions may be constructed from other functions.
- Functions have been built to construct pairs of values and select values from them.
- Normal order β reduction can be formalized in terms of substitution for free variables.
- α conversion is a way of removing name clashes in expressions.
- η reduction is a way of simplifying expressions.

Some of these topics are summarized below.

λ calculus syntax

<expression> ::= <name> | <function> | <application>

<name> ::= non-blank character sequence

<function> ::= λ <name> . <body>
<body> ::= <expression>

<application> ::= (<function expression>
<argument expression>)
<function expression> ::= <expression>
<argument expression> ::= <expression>

Free variables

- <name> is free in <name>.
- <name> is free in λ<name1>.<body>
if <name1> is not <name>
and <name> is free in <body>.
- <name> is free in (<function expression>
<argument expression>)
if <name> is free in <function expression>
or <name> is free in <argument expression>.

Bound variables

- <name> is bound in λ<name1>.<body>
  if <name> is <name1>
  or <name> is bound in <body>.
- <name> is bound in (<function expression>
  <argument expression>)
  if <name> is bound in <function expression>
  or <name> is bound in <argument expression>.

Normal order β reduction

For (<function expression> <argument expression>), containing no free variables:

(1) Normal order β reduce <function expression> to <function value>.

(2) If <function value> is λ<name>.<body>
then replace all free occurrences of <name> in <body> with
<argument expression>
and normal order β reduce the new <body>.

or

(3) If <function value> is not a function
then normal order β reduce <argument expression> to
<argument value>
and return (<function value> <argument value>).

Normal order reduction notation

=> – normal order β reduction
=> ... => – multiple normal order β reduction

Definitions

def <name> = <expression>

Replace all subsequent occurrences of <name> with <expression> before evaluation.

Replacement notation

== – defined name replacement

α conversion

To rename <name1> as <name2> in λ<name1>.<body>

if <name2> is not free in λ<name1>.<body>
then replace all free occurrences of <name1> in <body> with
<name2>
and replace <name1> in λ.<name1>

η reduction

(λ<name>.(<expression> <name>) <argument>) =>

<expression> <argument>

EXERCISES

2.1 Analyse each of the following λ expressions to clarify its structure. If the expression is a function, identify the bound variable and the body expression, and then analyse the body expression. If the expression is an application, identify the function and argument expressions, and then analyse the function and argument expressions:

(a) λa.(a λb.(b a))
(b) λx.λy.λz.((z x) (z y))
(c) (λf.λg.(λh.(g h) f) λp.λq.p)
(d) λfee.λfi.λfo.λfum.(fum (fo (fi fee)))
(e) (λp.(λq.p λx.(x p)) λi.λj.(j i))

2.2 Evaluate the following λ expressions:

(a) ((λx.λy.(y x) λp.λq.p) λi.i)
(b) (((λx.λy.λz.((x y) z) λf.λa.(f a)) λi.i) λj.j)
(c) (λh.((λa.λf.(f a) h) h) λf.(f f))
(d) ((λp.λq.(p q) (λx.x λa.λb.a)) λk.k)
(e) (((λf.λg.λx.(f (g x)) λs.(s s)) λa.λb.b) λx.λy.x)

2.3 For each of the following pairs, show that function (i) is equivalent to the function resulting from expression (ii) by applying both to arbitrary arguments:

(a) (i) identity
(ii) (apply (apply identity))
(b) (i) apply
(ii) λx.λy.(((make_pair x) y) identity)
(c) (i) identity
(ii) (self_apply (self_apply select_second))

2.4 Define a function:

```
def make_triplet = ...
```

which is like make_pair but constructs a triplet from a sequence of three arguments so that any one of the arguments may be selected by the subsequent application of a triplet to a selector function.

Define selector functions:

```
def triplet_first = ...
def triplet_second = ...
def triple_third = ...
```

which will select the first, second or third item from a triplet respectively.

Show that:

```
make_triplet <item1> <item2> <item3>
                                triplet_first => ... => <item1>
make_triplet <item1> <item2> <item3>
                                triplet_second => ... => <item2>
make_triplet <item1> <item2> <item3>
                                triplet_third => ... => <item3>
```

for the arbitrary arguments:

```
<item1> <item2> <item3>
```

2.5 Analyse each of the following λ expressions to identify its free and bound variables, and those in its subexpressions:

(a) λx.λy.(λx.y λy.x)
(b) λx.(x (λy.(λx.x y) x))
(c) λa.(λb.a λb.(λa.a b))
(d) (λfree.bound λbound.(λfree.free bound))
(e) λp.λq.(λr.(p (λq.(λp.(r q)))) (q p))

2.6 Use α conversion to ensure unique names in the expressions in Exercise 2.5 above.

Chapter 3
Conditions, booleans and numbers

In this chapter we are going to start to add layers to λ calculus to develop a higher level functional notation. Firstly, we will use the pair functions from Chapter 2 to represent conditional expressions with truth values true and false, and then use these to develop boolean operations like not, and and or. Next, we will use the pair functions to represent natural numbers in terms of the value zero and the successor function. Finally, notations for simplifying function definitions and λ expressions, and for an 'if ... then ... else' form of conditional expression, will be introduced. For the moment, we will be looking at untyped representations of truth values and functions. Typed representations will be developed in Chapter 5.

3.1 Truth values and conditional expression

Boolean logic is based on the truth values TRUE and FALSE with logical operations NOT, AND, OR and so on. We are going to represent TRUE by select_first and FALSE by select_second, and use a version of make_pair to build logical operations. To provide motivation, consider the conditional expression, in C:

```
<condition>?<expression>:<expression>
```

If the <condition> is TRUE, then the first <expression> is selected for evaluation and if the <condition> is FALSE, then the second <expression> is selected for evaluation. For example, to set max to the greater of x and y:

```
max = x>y?x:y
```

or to set absx to the absolute value of x:

```
absx = x<0?-x:x
```

We can model a conditional expression using a version of the make pair function:

```
def cond = λe1.λe2.λc.((c e1) e2)
```

Consider cond applied to the arbitrary expressions <expression1> and <expression2>:

```
((cond <expression1>) <expression2>) ==
((λe1.λe2.λc.((c e1) e2) <expression1>) <expression2>) =>
(λe2.λc.((c <expression1>) e2) <expression2>) =>
λc.((c <expression1>) <expression2>)
```

Now, if this function is applied to select_first:

```
(λc.((c <expression1>) <expression2>) select_first) =>
((select_first <expression1>) <expression2>) => ... =>
<expression1>
```

and if it is applied to select_second:

```
(λc.((c <expression1>) <expression2>) select_second) =>
((select_second <expression1>) <expression2>) => ... =>
<expression2>
```

Notice that the <condition> is the last argument for cond, not the first.

Now, we will use the conditional expression and cond function with:

```
def true = select_first

def false = select_second
```

to model some of the logical operators.

3.2 NOT

NOT is a unary operator of the form:

```
NOT <operand>
```

which we will describe through a truth table with X representing the single operand:

X	*NOT X*
FALSE	TRUE
TRUE	FALSE

Note that if the operand is TRUE, then the answer is FALSE and if the operand is FALSE, then the answer is TRUE. Thus NOT could be written using a conditional expression as:

```
X ? FALSE : TRUE
```

We can describe this using selectors so that if the operand is TRUE, then FALSE is selected and if the operand is FALSE, then TRUE is selected. This suggests using:

```
def not = λx.(((cond false) true) x)
```

Simplifying the inner body gives:

```
(((cond false) true) x) ==

(((λe1.λe2.λc.((c e1) e2) false) true) x) =>

((λe2.λc.((c false) e2) true) x) =>

(λc.((c false) true) x) =>

((x false) true)
```

so we will use:

```
def not = λx.((x false) true)
```

Let us try:

```
NOT TRUE
```

as:

```
(not true) ==
(λx.((x false) true) true) =>
((true false) true) ==
((λfirst.λsecond.first false) true) =>
(λsecond.false true) =>
false
```

and:

```
NOT FALSE
```

as:

```
(not false) ==
(λx.((x false) true) false) =>
((false false) true) ==
((λfirst.λsecond.second false) true) =>
(λsecond.second true) =>
true
```

which correspond to the truth table.

3.3 AND

AND is a binary operator of the form:

```
<operand> AND <operand>
```

which we will describe using a truth table with X representing the left operand and Y representing the right operand:

X	*Y*	*X AND Y*
FALSE	FALSE	FALSE
FALSE	TRUE	FALSE
TRUE	FALSE	FALSE
TRUE	TRUE	TRUE

Note that if the left operand is TRUE, then the final value depends on the right operand and if the left operand is FALSE, then the final value is FALSE so AND could be modelled using the conditional expression as:

```
X ? Y : FALSE
```

Using selectors, if the left operand is TRUE, then select the right operand and if the left operand is FALSE, then select FALSE, so we will define AND as:

```
def and = λx.λy.(((cond y) false) x)
```

Simplifying the inner body gives:

```
(((cond y) false) x) ==
(((λe1.λe2.λc.((c e1) e2) y) false) x) =>
((λe2.λc.((c y) e2) false) x) =>
(λc.((c y) false) x) =>
((x y) false)
```

so we will now use:

```
def and = λx.λy.((x y) false)
```

For example, we could write:

```
TRUE AND FALSE
```

as:

```
((and true) false) ==
((λx.λy.((x y) false) true) false) =>
(λy.((true y) false) false) =>
((true false) false) ==
```

```
((λfirst.λsecond.first false) false) =>

(λsecond.false false) =>

false
```

3.4 OR

OR is a binary operator of the form:

<operand> OR <operand>

which we will again describe with a truth table, using X for the left operand and Y for the right operand:

X	*Y*	*X OR Y*
FALSE	FALSE	FALSE
FALSE	TRUE	TRUE
TRUE	FALSE	TRUE
TRUE	TRUE	TRUE

Note that if the first operand is TRUE, then the final value is TRUE, otherwise the final value is the second operand. We could describe this using the conditional expression as:

```
X ? TRUE : Y
```

Using selectors, if the first operand is TRUE, then select TRUE, and if the first operand is FALSE then select the second operand:

```
def or = λx.λy.(((cond true) y) x)
```

Simplifying the inner body:

```
(((cond true) y) x) ==

(((λe1.λe2.λc.((c e1) e2) true) y) x) =>

((λe2.λc.((c true) e2) y) x) =>

(λc.((c true) y) x) =>

((x true) y)
```

Now we will use:

```
def or = λx.λy.((x true) y)
```

For example, we could write:

```
FALSE OR TRUE
```

as:

```
((or false) true) ==
((λx.λy.((x true) y) false) true) =>
(λy.((false true) y) true) =>
((false true) true) =>
((λfirst.λsecond.second true) true) =>
(λsecond.second true) =>
true
```

3.5 Natural numbers

We tend to take numbers for granted in programming but we now have to consider how to represent them explicitly. Our approach will be based on the ability to define **natural numbers** – non-negative integers – as **successors of zero:**

1 = successor of 0

2 = successor of 1
 = successor of successor of 0

3 = successor of 2
 = successor of successor of 1
 = successor of successor of successor of 0
 etc.

Thus, the definition of an arbitrary integer will be that number of successors of zero. We need to find a function zero to represent zero and a successor function succ so that we can define:

```
def one = (succ zero)
def two = (succ one)
def three = (succ two)
```

and so on. Note that:

```
two == (succ (succ zero))
three == (succ (succ one)) == (succ (succ (succ zero)))
```

and so on. There are a variety of ways of representing zero and succ. We will use:

```
def zero = identity
def succ = λn.λs.((s false) n)
```

so each time succ is applied to a number n it builds a pair function with false first and the original number second. For example:

```
one ==
(succ zero) ==
(λn.λs.((s false) n) zero) =>
λs.((s false) zero)
```

Similarly:

```
two ==
(succ one) ==
(λn.λs.((s false) n) one) =>
λs.((s false) one) ==
λs.((s false) λs.((s false) zero))
```

and:

```
three ==
(succ two) ==
(λn.λs.((s false) n) two) =>
λs.((s false) two) ==
λs.((s false) λs.((s false) one)) ==
λs.((s false) λs.((s false) λs.((s false) zero)))
```

This representation enables the definition of a unary function iszero which returns true if its argument is zero and false otherwise. Remember that a number is a function with an argument which may be used as a selector. For an arbitrary number:

```
λs.((s false) <number>)
```

if the argument is set to select_first then false will be selected:

```
(λs.((s false) <number>) select_first) =>
((select_first false) <number>) ==
((λfirst.λsecond.first false) <number>) =>
(λsecond.false <number>) =>
false
```

If zero, which is the identity function, is applied to select_first, then select_first, which is the same as true by definition, will be returned:

```
(zero select_first) ==
(λx.x select_first) =>
select_first ==
true
```

This suggests using:

```
def iszero = λn.(n select_first)
```

Notice that iszero applies the number to the selector rather than the selector to the number. This is because our number representation models numbers as functions with selector arguments.

We can now define the predecessor function pred so that:

```
(pred one) => ... => zero
(pred two) => ... => one
(pred three) => ... => two
```

and so on. For our number representation, pred should strip off a layer of nesting from an arbitrary number:

```
λs.((s false) <number>)
```

and return the:

```
<number>
```

This suggests using select_second because:

```
(λs.((s false) <number>) select_second) =>
((select_second false) <number>) ==
((λfirst.λsecond.second false) <number>) =>
(λsecond.second <number>) =>
<number>
```

so we might define a first version of pred as:

```
def pred1 = λn.(n select_second)
```

However, there is a problem with zero as we only have positive integers. Let us try our present pred1 with zero:

```
(pred1 zero) ==
(λn.(n select_second) zero) =>
(zero select_second) ==
(λx.x select_second) =>
select_second ==
false
```

which is not a representation of a number. We could define the predecessor of zero to be zero and check numbers to see if they are zero before returning their predecessor, using:

```
<number> = zero ? zero : predecessor of <number>
```

so:

```
def pred = λn.(((cond zero) (pred1 n)) (iszero n))
```

Simplifying the body gives:

```
(((cond zero) (pred1 n)) (iszero n)) ==
(((λe1.λe2.λc.((c e1) e2) zero) (pred1 n)) (iszero n)) =>
((λe2.λc.((c zero) e2) (pred1 n)) (iszero n)) =>
(λc.((c zero) (pred1 n)) (iszero n))  =>
(((iszero n) zero) (pred1 n))
```

Substituting for pred1 and simplifying gives:

```
(((iszero n) zero) (λn.(n select_second) n)) ==
(((iszero n) zero) (n select_second)) ==
```

so now we will use:

```
def pred = λn.(((iszero n) zero) (n select_second))
```

Alternatively, we might say that the predecessor of zero is undefined. We will not look at how to handle undefined values here. When we use pred we will have to be careful to check for a zero argument.

3.6 Simplified notations

By now you will have noticed that manipulating λ expressions involves lots of brackets. As well as being tedious and cumbersome, it is a major source of mistakes due to unmatched or mismatched brackets. To simplify things, we will allow brackets to be omitted when it is clear what is intended. In general, for the application of a function <function> to N arguments we will allow:

```
<function> <argument1> <argument2> ... <argumentN>
```

instead of:

```
(...((<function> <argument1>) <argument2>) ... <argumentN>)
```

so in a function application, a function is applied first to the nearest argument to the right. If an argument is itself a function application, then the brackets must stay. There must also be brackets around function body applications. For example, we could rewrite pred as:

```
def pred = λn.((iszero n) n (n select_second))
```

We can also simplify name/function association definitions by dropping the 'λ' and '.', and moving the bound variable to the left of the '=' so:

```
def <names> = λ<name>.<expression>
```

where <names> is one or more <name>s, becomes:

```
def <names> <name> = <expression>
```

We can now rewrite all of our definitions:

```
def identity x = x

def self_apply s = s s

def apply func = λarg.(func arg)
```

and hence:

```
    def apply func arg = func arg

def select_first first = λsecond.first
```

and hence:

```
    def select_first first second = first

def select_second first = λsecond.second
```

and hence:

```
    def select_second first second = second

def make_pair e1 = λe2.λc.(c e1 e2)
```

and hence:

```
def make_pair e1 e2 = λc.(c e1 e2)
```

and hence:

```
    def make_pair e1 e2 c = c e1 e2

def cond e1 e2 c = c e1 e2

def true first second = first

def false first second = second

def not x = x false true

def and x y = x y false

def or x y = x true y
```

For some functions there are standard equivalent notations. Thus, it is usual to write:

```
cond <true choice> <false choice> <condition>
```

in an if ... then ... else ... form. We will use:

```
if <condition>
then <true choice>
else <false choice>
```

For example, we could rewrite pred's definition as:

```
def pred n =
 if iszero n
 then zero
 else n select_second
```

Similarly, using our conditional derivation of booleans, we could rewrite not as:

```
def not x =
 if x
 then false
 else true
```

and and as:

```
def and x y =
 if x
 then y
 else false
```

and or as:

```
def or x y =
 if x
 then true
 else y
```

SUMMARY

- Conditional expressions with truth values may be represented by pair functions and used to develop boolean operations.
- The definition of natural numbers may be based on zero and the successor function.
- Notations for removing brackets from expressions, simplifying function definitions and an 'if ... then ... else ...' form of conditional expression are summarized below.

Removing brackets

```
( ... ((<function> <argument1>) <argument2>)
      ... <argumentN>) ==

<function> <argument1> <argument2> ... <argumentN>
```

Simplifying function definitions

```
def <names> = λ<name>.<expression> ==

def <names> <name> = <expression>
```

if ... then ... else ...

```
if <condition>
then <true choice>
else <false choice> ==

cond <true choice> <false choice> <condition>
```

EXERCISES

3.1 The boolean operation implication is defined by the following truth table:

X	*Y*	*X IMPLIES Y*
FALSE	FALSE	TRUE
FALSE	TRUE	TRUE
TRUE	FALSE	FALSE
TRUE	TRUE	TRUE

Define a λ calculus representation for implication:

```
def implies = λx.λy...
```

Show that the definition satisfies the truth table for all boolean values of x and y.

3.2 The boolean operation equivalence is defined by the following truth table:

X	*Y*	*X EQUIV Y*
FALSE	FALSE	TRUE
FALSE	TRUE	FALSE
TRUE	FALSE	FALSE
TRUE	TRUE	TRUE

Define a λ calculus representation for equivalence:

```
def equiv = λx.λy...
```

Show that the definition satisfies the truth table for all boolean values of x and y.

3.3 For each of the following pairs, show that functions (i) and (ii) are equivalent for all boolean values of their arguments:

(a) (i) λx.λy.(and (not x) (not y))
(ii) λx.λy.(not (or x y))

(b) (i) implies
(ii) λx.λy.(implies (not y) (not x))

(c) (i) not
(ii) λx.(not (not (not x)))

(d) (i) implies
(ii) λx.λy.(not (and x (not y)))

(e) (i) equiv
(ii) λx.λy.(and (implies x y) (implies y x))

3.4 Show that:

```
λx.(succ (pred x))
```

and:

```
λx.(pred (succ x))
```

are equivalent for arbitrary non-zero integer arguments. Explain why they are not equivalent for a zero argument.

Chapter 4
Recursion and arithmetic

In this chapter, we are going to look at how recursion is used for repetition in functional programming. To begin with, we will see that our existing definition notation cannot be used to introduce recursion because it leads to infinite substitution sequences. We will then show that this can be overcome through abstraction in individual functions. Recursion will be introduced using a general purpose construct based on function self-application, and we will look at its use in building a wide variety of arithmetic operations.

4.1 Repetition, iteration and recursion

Repetition involves doing the same thing zero or more times. It is useful to distinguish **bounded repetition,** where something is carried out a fixed number of times, from the more general **unbounded iteration,** where something is carried out until some condition is met. Thus, for bounded repetition the number of repetitions is known in advance whereas for unbounded repetition it is not.

It is important to relate the form of repetition to the structure of the item to be processed. Bounded repetition is used where a **linear** sequence of objects of known length is to be processed, for example, processing each element of an array. Here, the object sequence can be related to a consecutive range of values, since arrays have addresses which are linear sequences of integers.

Unbounded repetition is used where a **nested** sequence of objects is to be processed and the number of layers of nesting is unknown. For example, a filing system might consist of a nested hierarchy of directories and files. Processing such a filing system involves starting at the root directory and then processing the files and subdirectories. Processing the subdirectories involves processing their files and subdirectories, and so on. In general, the depth of directory nesting is unknown. For unbounded repetition, processing is complete when the end of the nesting is reached. For example, processing a filing system ends when all the files at every level of directory nesting have been processed.

Bounded repetition is a weaker form of unbounded repetition. Carrying out something a fixed number of times is the same as carrying it out until the last item in a sequence has been dealt with.

In imperative languages, repetition is based primarily on **iterative** constructs for repeatedly carrying out structured assignment sequences. For example, in Pascal, FOR statements provide bounded iteration over a range of integers and WHILE or REPEAT statements provide unbounded iteration until a condition is met. Here, repetition involves repeatedly inspecting and changing variables in common memory.

In functional languages, programs are based on structured nested function calls. Repetition requires such nesting to be continued until some condition is met. There is no concept of a changing shared memory. Instead, each function passes its result to the next in the function call nesting. Repetition involves deciding whether or not to carry out another layer of function call nesting with the same nested sequence of function calls.

Repetition in functional programming is based on **recursion**: the definition of something in terms of itself. The nested function call sequence which is to be repeated is given a name. If some condition is met within the sequence, then that name is invoked to repeat the nested function call sequence within itself. The condition often checks whether or not the end of a linear or nested object sequence has been reached.

Let us compare iteration and recursion through another contrived example. Suppose we want to eat some apples. If we know that there are N apples then we might write an iterative algorithm as:

```
EAT N = FOR COUNT := N DOWNTO 1 DO
         gobble an apple
```

or:

```
EAT N = COUNT := N
        WHILE COUNT > 0 DO
        BEGIN
             gobble an apple
             COUNT := COUNT - 1
        END
```

For example, for 3 apples we would:

```
EAT 3 apples =>

gobble an apple and
gobble an apple and
gobble an apple and
stop
```

An equivalent recursive algorithm would be:

```
EAT N = IF N > 0 THEN
        BEGIN
             gobble an apple
             EAT N-1
        END
```

For example, for 3 apples we would:

```
EAT 3 apples =>

gobble an apple and
EAT 2 apples =>

gobble an apple and
gobble an apple and
EAT 1 apple =>

gobble an apple and
gobble an apple and
gobble an apple and
EAT 0 apples =>
```

```
gobble an apple and
gobble an apple and
gobble an apple and
stop
```

Note that eating apples iteratively involves gobbling 1 apple N times, whereas eating apples recursively involves gobbling 1 apple and then eating the remaining N−1 apples recursively.

It is useful to distinguish **primitive recursion**, where the number of repetitions is known, from **general recursion**, where the number of repetitions is unknown. Primitive recursion is weaker than general recursion. Primitive recursion involves a finite depth of function call nesting, so it is equivalent to bounded repetition through iteration with a finite memory. For general recursion, the nesting depth is unknown so it is equivalent to unbounded repetition through iteration with an infinite memory.

Note that imperative languages often provide repetition through recursive procedures and functions as well as through iteration.

4.2 Recursion through definitions?

It might appear that our definition notation enables recursion and we could use the name from the left of the definition in the expression on the right. For example, two numbers may be added together by repeatedly incrementing the first and decrementing the second until the second is zero:

```
def add x y =
 if iszero y
 then x
 else add (succ x) (pred y)
```

Thus, to add one and two, for example:

```
add one two => ... =>

add (succ one) (pred two) => ... =>

add (succ (succ one)) (pred (pred two)) => ... =>

(succ (succ one)) ==

three
```

However, in Chapter 2 we required *all names in expressions to be replaced by their definitions before the expression is evaluated*. In the above example:

```
λx.λy.
 if iszero y
 then x
 else add (succ x) (pred y) ==

λx.λy.
 if iszero y
 then x
 else
  ((λx.λy.
    if iszero y
    then x
    else add (succ x) (pred y))
   (succ x) (pred y)) ==

λx.λy.
 if iszero y
 then x
 else
  ((λx.λy.
   if iszero y
   then x
   else
    ((λx.λy.
      if iszero y
      then x
      else add (succ x) (pred y))
     (succ x) (pred y))
   (succ x) (pred y)) == ...
```

Replacement will never terminate!

We want the replacement to take place a finite number of times, depending on particular uses of the function with particular arguments but, of course, there is no way of knowing what argument values are required when the function is defined. If we did know, then we could construct specific functions for specific cases, rather than a general purpose function. This was not a problem in earlier examples, because we knew replacement would always be finite. For recursion, though, we need some means of delaying the repetitive use of the function until it is actually required.

4.3 Passing a function to itself

Function use always occurs in an application and may be delayed through abstraction at the point where the function is used. For an arbitrary function, the application:

```
<function> <argument>
```

is equivalent to:

```
λf.(f <argument>) <function>
```

The original function becomes the argument in a new application.

In our addition example, we could introduce a new argument:

```
def add1 f x y =
 if iszero y
 then x
 else f (succ x) (pred y)
```

to remove recursion by abstraction at the point where recursion is required. Now we need to find an argument for add1 with the same effect as add. Of course, we cannot just pass add to add1 as we end up with the non-terminating replacement again. What is needed is to pass add1 into itself, but this just pushes the problem down a level. If we try:

```
def add = add1 add1
```

then the definition expands to:

```
(λf.λx.λy.
   if iszero y
   then x
   else f (succ x) (pred y)) add1 =>
λx.λy.
 if iszero y
 then x
 else add1 (succ x) (pred y)
```

We have failed to pass add1 down far enough. In the original definition for add1, the application:

```
f (succ x) (pred y)
```

has only two arguments. Thus, after substitution:

```
add1 (succ x) (pred y)
```

has no argument corresponding to the bound variable f. We need the effect of:

```
add1 add1 (succ x) (pred y)
```

so that add1 may be passed on to subsequent recursions.

Let us define an add2, this time passing the argument for f to the argument itself as well:

```
def add2 f x y =
 if iszero y
 then x
 else f f x y
```

As before, add is:

```
def add = add2 add2
```

The definition expands and evaluates as:

```
(λf.λx.λy.
  if iszero y
  then x
  else f f (succ x) (pred y)) add2 =>

λx.λy.
 if iszero y
 then x
 else add2 add2 (succ x) (pred y)
```

Note that we do not strictly need to replace other occurrences of add2, as its definition contains no references to itself.

Now, we have inserted two copies of add2 – one as function and another as argument – to continue recursion. Thus, every time the recursion point is reached, another copy of the whole function is passed down. For example:

```
add one two ==

(λx.λy.
  if iszero y
  then x
  else add2 add2 (succ x) (pred y)) one two  => ... =>

if iszero two
then one
else add2 add2 (succ one) (pred two)  => ... =>

(λf.λx.λy.
  if iszero y
```

```
    then x
    else f f (succ x) (pred y)) add2 (succ one) (pred two)
     => ... =>

  if iszero (pred two)
  then (succ one)
  else add2 add2 (succ (succ one)) (pred (pred two)) => ... =>

  (λf.λx.λy.
    if iszero y
    then x
    else f f (succ x) (pred y)) add2 (succ (succ one))
                                     (pred (pred two))  => ... =>

  if iszero (pred (pred two))
  then (succ (succ one))
  else add2 add2 (succ (succ (succ one)))
                 (pred (pred (pred two))) => ... =>

  succ (succ one)) ==

  three
```

4.4 Applicative order reduction

From now on, to simplify the presentation of some examples, we will evaluate them partially in applicative order, that is, in some cases we will evaluate arguments before passing them to functions. We will indicate the applicative order reduction of an argument with:

```
->
```

and the applicative order reduction of a sequence of arguments with:

```
-> ... ->
```

Note that argument evaluation will generally involve other reductions which will not be shown.

We will consider the relationship between applicative and normal order reduction in Chapter 8, but note now that the result of a terminating applicative order reduction of an expression is the same as the result of the equivalent terminating normal order reduction. As we will see in Chapter 8, the reverse is not true because there are expressions with terminating normal order reductions but non-terminating applicative order reductions. Nonetheless, provided evaluation terminates, applicative and normal order are equivalent. As we will also see in Chapter 8, a major source of non-termination results from our representation of conditional expressions. It turns out that the strict applicative order reduction of conditional expressions embodying recursive calls in a function body will not terminate. Thus, until Chapter 8, the use of the applicative order indicators:

```
->
```

and:

```
-> ... ->
```

will still imply the normal order reduction of conditional expressions.

4.5 Recursion function

A more general approach to recursion is to find a constructor function to build a recursive function from a non-recursive function, with a single abstraction at the recursion point. For example, we might define multiplication recursively. To multiply two numbers, add the first to the product of the first and the decremented second. If the second is zero, then so is the product:

```
def mult x y =
 if iszero y
 then zero
 else add x (mult x (pred y))
```

For example:

```
mult three two => ... =>

add three (mult three (pred two)) -> ... ->

add three (add three (mult three (pred (pred two)))) -> ... ->

add three (add three zero) -> ... ->

add three three => ... =>

six
```

We can remove self-reference by abstraction at the recursion point:

```
def mult1 f x y =
 if iszero y
 then zero
 else add x (f x (pred y))
```

We would like to have a function recursive, which will construct recursive functions from non-recursive versions, for example:

```
def mult = recursive mult1
```

The function recursive must not only pass a copy of its argument to that argument but also ensure that self-application will continue: the copying mechanism must be passed on as well. This suggests that recursive should be of the form:

```
def recursive f = f <'f' and copy>
```

If recursive is applied to mult1:

```
recursive mult1 ==

λf.(f <'f' and copy>) mult1 =>

mult1 <'mult1' and copy> ==

(λf.λx.λy.
  if iszero y
  then zero
  else add x (f x (pred y))) <'mult1' and copy> =>

λx.λy.
 if iszero y
 then zero
 else add x (<'mult1' and copy> x (pred y))
```

In the body we have:

```
<'mult1' and copy> x (pred y)
```

but we require:

```
mult1 <'mult1' and copy> x (pred y)
```

so that:

```
<'mult1' and copy>
```

is passed on again through mult1's bound variable f to the next level of recursion. Thus, the copy mechanism must be such that:

```
<'mult1' and copy> => ... => mult1 <'mult1' and copy>
```

In general, from function f passed to recursive, we need:

```
<'f' and copy> => ... => f <'f' and copy>
```

so the copy mechanism must be an application and that application must be self-replicating. We know that the self-application function:

```
λs.(s s)
```

will self-replicate when applied to itself but the replication never ends. Self-application may be delayed through abstraction with the construction of a new function:

```
λf.λs.(f (s s))
```

Here, the self-application:

```
(s s)
```

becomes an argument for f. This might, for example, be a function with a conditional expression in its body, which will only lead to the evaluation of its argument when some condition is met. When this new function is applied to an arbitrary function, we get:

```
λf.λs.(f (s s)) <function> =>
λs.(<function> (s s))
```

If this function is now applied to itself:

```
λs.(<function> (s s)) λs.(<function> (s s)) =>
<function> (λs.(<function> (s s)) λs.(<function> (s s)))
```

then we have a copy mechanism which matches our requirement.

Thus, we can define recursive as:

```
def recursive f = λs.(f (s s)) λs.(f (s s))
```

For example, in:

```
def mult = recursive mult1
```

the definition evaluates to:

```
λf.(λs.(f (s s)) λs.(f (s s))) mult1 =>
λs.(mult1 (s s)) λs.(mult1 (s s)) =>
```

```
mult1 (λs.(mult1 (s s)) λs.(mult1 (s s))) ==

(λf.λx.λy.
  if iszero y
  then zero
  else add x (f x (pred y))) (λs.(mult1 (s s)) λs.(mult1 (s s))) =>

λx.λy.
 if iszero y
 then zero
 else add x ((λs.(mult1 (s s)) λs.(mult1 (s s)))
            x (pred y))
```

Again, note that we do not strictly need to replace other occurrences of mult1, since its definition contains no references to itself. For example, we will try:

```
mult three two => ... =>

(λx.λy.
  if iszero y
  then zero
  else add x ((λs.(mult1 (s s)) λs.(mult1 (s s)))
             x (pred y))) three two   => ... =>

if iszero two
then zero
else add three ((λs.(mult1 (s s)) λs.(mult1 (s s)))
               three (pred two))   => ... =>

add three ((λs.(mult1 (s s)) λs.(mult1 (s s)))
          three (pred two)) ->

add three (mult1 (λs.(mult1 (s s)) λs.(mult1 (s s)))
                 three (pred two)) ==

add three ((λx.λy.
             if iszero y
             then zero
             else add x ((λs.(mult1 (s s)) λs.(mult1 (s s)))
                         x (pred y))) three (pred two))   -> ...->

add three if iszero (pred two)
          then zero
          else add three ((λs.(mult1 (s s)) λs.(mult1 (s s)))
                          three (pred (pred two)))   -> ... ->

add three (add three ((λs.(mult1 (s s)) λs.(mult1 (s s)))
                     three (pred (pred two)))) ->
```

```
add three (add three (mult1 (λs.(mult1 (s s)) λs.(mult1 (s s)))
                                     three (pred (pred two)))) ==

add three (add three ((λx.λy.
                        if iszero y
                        then zero
                        else add x ((λs.(mult1 (s s)) λs.(mult1 (s s)))
                                   x (pred y))
                      three (pred (pred two))))  -> ... ->

add three (add three if iszero (pred (pred two))
                     then zero
                     else add three
                        ((λs.(mult1 (s s)) λs.(mult1 (s s)))
                        three (pred (pred (pred two))))) -> ... ->

add three (add three zero) -> ... ->

add three three => ... =>

six
```

4.6 Recursion notation

The function recursive is known as a **paradoxical combinator** or a **fixed point finder,** and is called **Y** in λ calculus literature. Rather than always defining an auxiliary function with an abstraction and then using recursive to construct a recursive version, we will allow the defined name to appear in the defining expression but use a new definition form:

```
rec <name> = <expression>
```

This indicates that the occurrence of the name in the definition should be replaced using abstraction, and the paradoxical combinator should then be applied to the whole of the defining expression. For example, for addition, we will write:

```
rec add x y =
 if iszero y
 then x
 else add (succ x) (pred y)
```

instead of:

```
def add1 f x y =
 if iszero y
```

```
 then x
 else f (succ x) (pred y)

def add = recursive add1
```

and for multiplication we will write:

```
rec mult x y =
 if iszero y
 then zero
 else add x (mult x (pred y))
```

When we expand or evaluate a recursive definition we will leave the recursive reference in place.

4.7 Arithmetic operations

We will now use recursion to define arithmetic operations: raising to a power, subtraction, equality and inequalities, and division.

4.7.1 Raising to a power

To raise one number to the power of another number, multiply the first by the first to the power of the decremented second. If the second is zero, then the power is one:

```
rec power x y =
 if iszero y
 then one
 else mult x (power x (pred y))
```

For example:

```
power two three => ... =>

mult two
     (power two (pred three)) -> ... ->

mult two
     (mult two
           (power two (pred (pred three)))) -> ... ->

mult two
     (mult two
           (mult two
```

```
                 (power two (pred (pred(pred three))))))
                  -> ... ->
mult two
     (mult two
          (mult two one)) -> ... ->
mult two
     (mult two two) -> ... ->
mult two four => ... =>
eight
```

4.7.2 Subtraction

To find the difference between two numbers, find the difference between the numbers after decrementing both. The difference between a number and zero is the number:

```
rec sub x y =
 if iszero y
 then x
 else sub (pred x) (pred y)
```

For example:

```
sub four two => ... =>
sub (pred four) (pred two) => ... =>
sub (pred (pred four)) (pred (pred two)) => ... =>
(pred (pred four)) => ... =>
two
```

Notice that this version of subtraction will return zero if the second number is larger than the first, for example:

```
sub one two => ... =>
sub (pred one) (pred two) => ... =>
sub (pred (pred one)) (pred (pred two)) => ... =>
pred (pred one) -> ... ->
pred zero => ... =>
zero
```

This is because pred returns zero from decrementing zero. This form of subtraction is known as **natural subtraction.**

4.7.3 Comparison

There are a number of ways of defining equality between numbers. One approach is to notice that the difference between two equal numbers is zero. However, if we subtract a number from a smaller number we also get zero, so we need to find the **absolute difference** between them; the difference regardless of the order of comparison. To find the absolute difference between two numbers, add the difference between the first and the second to the difference between the second and the first:

```
def abs_diff x y = add (sub x y) (sub y x)
```

If they are both the same, then the absolute difference will be zero because the result of taking each from the other will be zero. If the first is greater than the second, then the absolute difference will be the first minus the second because the second minus the first will be zero. Similarly, if the second is greater than the first, then the difference will be the second minus the first because the first minus the second will be zero. Thus, we can define:

```
def equal x y = iszero (abs_diff x y)
```

For example:

```
equal two three => ... =>

iszero (abs_diff two three) -> ... ->

iszero (add (sub two three) (sub three two)) -> ... ->

iszero (add zero one) -> ... ->

iszero one => ... =>

false
```

We could equally well be explicit about the decrementing sub carries out and define equality recursively. Two numbers are equal if both are zero; they are unequal if one is zero or equal if decrementing both gives equal numbers:

```
rec equal x y =
 if and (iszero x) (iszero y)
 then true
```

```
else
  if or (iszero x) (iszero y)
  then false
  else equal (pred x) (pred y)
```

For example:

```
equal two two => ... =>
equal (pred two) (pred two) -> ... ->
equal one one => ... =>
equal (pred one) (pred one) -> ... ->
equal zero zero => ... =>
true
```

We can also use subtraction to define arithmetic inequalities. For example, a number is greater than another if subtracting the second from the first gives a non-zero result:

```
def greater x y = not (iszero (sub x y))
```

For example, for:

```
3 > 2
```

we use:

```
greater three two => ... =>
not (iszero (sub three two)) -> ... ->
not (iszero one) -> ... ->
not false => ... =>
true
```

Similarly, a number is greater than or equal to another if taking the first from the second gives zero:

```
def greater_or_equal x y = iszero (sub y x)
```

For example, for:

```
2 >= 3
```

we use:

```
greater_or_equal two three => ... =>

iszero (sub three two) -> ... ->

iszero one => ... =>

false
```

4.7.4 Division

Division, like decrementation, is problematic because of zero. It is usual to define division by zero as undefined, but we do not have any way of dealing with undefined values. Let us define division by zero to be zero and remember to check for a zero divisor. For a non-zero divisor, we count how often it can be subtracted from the dividend until the dividend is smaller than the divisor:

```
rec div1 x y =
 if greater y x
 then zero
 else succ (div1 (sub x y) y)

def div x y =
 if iszero y
 then zero
 else div1 x y
```

For example:

```
div nine four => ... =>

div1 nine four => ... =>

succ (div1 (sub nine four) four)) -> ... ->

succ (div1 five four) -> ... ->

succ (succ (div1 (sub five four) four)) -> ... ->

succ (succ (div1 one four)) -> ... ->

succ (succ zero) -> ... ->

two
```

SUMMARY

- Recursion is a means of repetition in functional programming.
- Recursion through function definitions leads to non-terminating substitution sequences.

- Recursion may be enabled by abstracting at the place where recursion takes place in a function and then passing the function to itself.
- Evaluation may be simplified by applicative order β reduction.
- Recursion may be generalized through a recursion function which substitutes a function at its own recursion points.
- Recursive functions may be defined by recursion notation.
- Recursion can be used to develop standard arithmetic operations.

Some of these topics are summarized below.

Recursion by passing a function to itself

For:

```
def <name> = ... (<name> ...) ...
```

write:

```
def <name1> f  = ... (f f ... ) ...

def <name> = <name1> <name1>
```

Applicative order β reduction

For (<function expression> <argument expression>) containing no free variables:

- applicative order β reduce <argument expression> to <argument value>
- applicative order β reduce <function expression> to <function value>
- if <function value> is λ<name>.<body>
  then replace all free occurrences of <name> in <body> with <argument value>
  and applicative order β reduce the new <body>

or

- if <function value> is not a function
  then return (<function value> <argument value>)

Applicative order reduction notation

-> – applicative order β reduction

-> ... -> – multiple applicative order β reduction

Recursion function

```
def recursive f = λs.(f (s s)) λs.(f (s s))
```

For:

```
def <name> = ... (<name> ... ) ...
```

write:

```
def <name1> f = ... (f ... ) ...

def <name> = recursive <name1>
```

Note that:

```
recursive <name1> => ... => <name1> (recursive <name1>)
```

Recursion notation

```
rec <name> = <expression using '<name>'> ==

def <name> = recursive λf.<expression using 'f'>
```

EXERCISES

4.1 The following function finds the sum of the numbers between n and zero:

```
def sum1 f n =
 if iszero n
 then zero
 else add n (f (pred n))

def sum = recursive sum1
```

Evaluate:

```
sum three
```

4.2 Write a function that finds the product of the numbers between n and one:

```
def prod1 f n = ...

def prod = recursive prod1
```

so that:

```
prod n
```

in λ calculus is equivalent to:

```
n * n−1 * n−2 * ... * 1
```

in normal arithmetic (there is no escape from 'factorial' ...). Evaluate:

```
prod three
```

4.3 Write a function which finds the sum of applying a function fun to the numbers between n and zero:

```
def fun_sum1 f fun n = ...
def fun_sum = recursive fun_sum1
```

For example, given the 'squaring' function:

```
def sq x = mult x x
```

then:

```
fun_sum sq three
```

in λ calculus is equivalent to:

$$3^2 + 2^2 + 1^2 + 0^2$$

in arithmetic. Evaluate:

```
fun_sum double three
```

given the 'doubling' function:

```
def double x = add x x
```

4.4 Define a function to find the sum of applying a function fun to the numbers between n and zero in steps of s:

```
def fun_sum_step1 f fun n s = ...
def fun_sum_step = recursive fun_sum_step1
```

so, for example:

```
fun_sum_step sq six two
```

in λ calculus is equivalent to:

$$6^2 + 4^2 + 2^2 + 0^2$$

in normal arithmetic. Evaluate:

(a) fun_sum_step double five two
(b) fun_sum_step double four two

4.5 Define functions to test whether or not a number is less than, or less than or equal to another number:

```
def less x y = ...

def less_or_equal x y = ...
```

Evaluate:

(a) less three two
(b) less two three
(c) less two two
(d) less_or_equal three two
(e) less_or_equal two three
(f) less_or_equal two two

4.6 Define a function to find the remainder on dividing one number by another:

```
def mod x y = ...
```

Evaluate:

(a) mod three two
(b) mod two three
(c) mod three zero

Chapter 5
Types

In this chapter we are going to consider how types can be added to our functional notation to ensure that only meaningful arguments are passed to functions. We will consider the role of types in programming in general and how types may be characterized. We will then introduce functions for constructing and manipulating typed values, using the pair manipulation functions to represent typed objects as type/value pairs. The error type for error objects which are returned after type errors will be introduced, and typed representations for booleans, numbers and characters will be developed. Finally, new notations will be used to simplify function definitions through case definitions and structure matching.

5.1 Types and programming

We are working with a very simple language. As we exclude single names as expressions, the only objects are functions which take function arguments and return function results. (For the moment, we will not consider non-terminating applications.) We have constructed functions which we can interpret as boolean values, boolean operations, numbers, arithmetic operations and so on, but particular functions have no intrinsic interpretations other than in terms of their effects on other functions. Because functions are so general, there is no way to restrict the application of functions to other specific functions, for example, we cannot restrict 'arithmetic' functions to 'numeric' operands. We can carry out function applications which are perfectly valid, but have results with no relevant meaning within our intended interpretations. For example, consider the effect of:

```
iszero true ==
λn.(n select_first) true =>
true select_first ==
λfirst.λsecond.first select_first =>
λsecond.select_first ==
λsecond.λfirst.λsecond.first
```

This produces a perfectly good function for selecting the second argument in an application with three nested arguments, but we expect iszero to return a boolean.

Using these functions is analogous to programming in machine code. In most Central Processing Units (CPUs), the sole objects are undifferentiated bit patterns which have no intrinsic meanings but are interpreted in different ways by the machine code operations applied to them. For example, different machine code instructions may treat a bit pattern as a signed or unsigned integer, decimal or floating point number or a data or instruction address. Thus, a machine code program may two's-complement an address or jump to a floating point number.

The single-typed systems programming language BCPL, a precursor of C, is similarly free and easy. Although representations are provided for a variety of objects, operations are used without type checks on the assumption that their operands are appropriate. Early versions of C provided type checking, but were somewhat lax when operations were carried out on almost appropriate types, allowing indirection on integers or arithmetic on pointers, for example.

It is claimed that this 'freedom' from types makes languages more flexible. It does ease implementation dependent programming, where advantage is taken of particular architectural features on particular CPUs through bit or word level manipulation, but this in turn leads to a loss of portability because of gross incompatibilities between architectures. For example, many computer memories are based on 8-bit bytes so 16-bit words require 2 bytes. However, computers differ in the order in which these bytes are used: some put the top 8 bits in the first byte but others put them in the second byte. Thus, programs using 'clever' address arithmetic which involves knowing the byte order will not work on some computers. 'Type free' programming also increases inexplicable errors through questionable low-level subterfuges. For example, 'cunning' address manipulations to access the fields of a data structure may cause the corruption of other fields, or of a completely different data structure which is close to the requisite one in memory, through byte misalignments.

5.2 Types as objects and operations

Types are introduced into languages to control the use of operations on objects in order to ensure that only meaningful combinations are used. As we saw in Chapter 2, variables in programming languages are used as a primary abstraction mechanism. In 'typeless' languages there are no restrictions on object/operation combinations and any variable may be associated with any object, that is, variables can abstract over objects in general. In weakly typed languages, like LISP and PROLOG, objects are typed but variables are not. There are restrictions on object/operation combinations but not on variable/object associations. Thus, variables do not abstract over specific types. In strongly typed languages, like ML and Pascal, variables are specified as being of a particular type and have the same restrictions on use as objects of that type.

Formally, a type specifies a class of objects and associated operations. Object classes may be defined by listing their values, for example, for booleans:

```
TRUE is a boolean
FALSE is a boolean
```

or by specifying a base object and a means of constructing new objects from the base, for example, for natural numbers:

```
0 is a number

SUCC N is a number
if N is a number
```

Thus, we can show that:

```
SUCC (SUCC (SUCC 0))
```

is a number because:

```
0
SUCC 0
SUCC (SUCC 0)
```

are all numbers.

Operations may be specified exhaustively with a case for each base value. For example, for booleans, negation:

```
NOT TRUE = FALSE
NOT FALSE = TRUE
```

and conjunction:

```
AND FALSE FALSE = FALSE
AND FALSE TRUE = FALSE
AND TRUE FALSE = FALSE
AND TRUE TRUE = TRUE
```

Operations may also be specified constructively in terms of base cases for the base objects and general cases for the constructive objects. For example, for natural numbers, the predecessor function:

```
PRED 0 = 0
PRED (SUCC X) = X
```

and addition:

```
ADD X 0 = X
ADD X (SUCC Y) = ADD (SUCC X) Y
```

and subtraction:

```
SUB X 0 = X
SUB (SUCC X) (SUCC Y) = SUB X Y
```

and multiplication:

```
MULT X 0 = 0
MULT X (SUCC Y) = ADD X (MULT X Y)
```

Note that we have just formalized the informal descriptions from the last chapter. We will look at the relationship between exhaustive and case definitions, and our conditional style of definition later in this chapter.

Sometimes, it may be necessary to introduce conditions into case definitions, because the form of the object definition may not provide enough information to discriminate between cases. For example, for division:

```
DIV X 0 = NUMBER ERROR
DIV X Y = 0
          if (GREATER Y X)
DIV X Y = SUCC (DIV (SUB X Y) Y)
          if NOT (GREATER Y X)
```

the ifs are needed because the *values* rather than the *structures* of X and Y determine how they are to be processed.

Operations may map objects of a type to objects of the same type or to objects of another type. A common requirement is for **predicates** which are used to test properties of objects and return booleans. For example, for numbers:

```
EQUAL 0 0 = TRUE
EQUAL (SUCC X) 0 = FALSE
EQUAL 0 (SUCC X) = FALSE
EQUAL (SUCC X) (SUCC Y) = EQUAL X Y
```

We are not going to give a full formal treatment of types here.

5.3 Representing typed objects

We are going to construct functions to represent typed objects. In general, an object will have a type and a value. We need to be able to:

(1) construct an object from a value and a type
(2) select the value and type from an object
(3) test the type of an object
(4) handle type errors

We will represent an object as a type/value pair:

```
def make_obj type value = λs.(s type value)
```

An arbitrary object of type:

```
<type>
```

and value:

```
<value>
```

is represented by:

```
make_obj <type> <value> => ... =>
λs.(s <type> <value>)
```

Thus, the type is selected with select_first:

```
def type obj = obj select_first
```

and the value is selected with select_second:

```
def value obj = obj select_second
```

We will use numbers to represent types and numeric comparison to test the type:

```
def istype t obj = equal (type obj) t
```

We are going to define typed objects and operations in terms of untyped objects and operations. In general, our approach will be:

(1) check argument types
(2) extract untyped values from typed arguments
(3) carry out untyped operations on untyped values
(4) construct typed result from untyped result

We must distinguish definitions from the subsequent uses of typed objects. When defining types, we cannot avoid using untyped operations: we have to start somewhere. Once types are defined, however, we should only manipulate typed objects with typed operations to ensure that the type checks are not overridden.

In general, we will use UPPER CASE LETTERS for typed constructs and lower case letters for untyped constructs. We should show that our representation of a type satisfies the formal definition of that type but we will not do this rigorously or religiously.

5.4 Errors

Whenever we detect a type error we will return an appropriate error object. Such an object will have type error_type, represented as zero:

```
def error_type = zero
```

We need a function to construct error objects:

```
def MAKE_ERROR = make_obj error_type
```

This definition expands as:

```
make_obj error_type ==
λtype.λvalue.λs.(s type value) error_type =>
λvalue.λs.(s error_type value)
```

An error object's value should indicate the sort of error the object represents. For example, for a type error the corresponding error object might have the expected type as its value.

We will define a universal error object of type error_type:

```
def ERROR = MAKE_ERROR error_type
```

This definition expands as:

```
λvalue.λs.(s error_type value) error_type =>
λs.(s error_type error_type)
```

so the error object ERROR has type error_type and value error_type.

We can test for an object of type error by using istype to look for error_type:

```
def iserror = istype error_type
```

so iserror's definition expands as:

```
istype error_type ==
λt.λobj.(equal (type obj) t) error_type =>
λobj.(equal (type obj) error_type)
```

For example, to test that ERROR is of type error:

```
iserror ERROR ==
λobj.(equal (type obj) error_type) ERROR =>
equal (type ERROR) error_type
```

Now:

```
type ERROR
```

expands as:

```
λobj.(obj select_first) ERROR =>
ERROR select_first ==
λs.(s error_type error_type) select_first =>
select_first error_type error_type => ... =>
errortype
```

Thus:

```
equal (type ERROR) error_type -> ... ->
equal error_type error_type => ... =>
true
```

Our formal type definitions should be extended to show how error objects are accommodated, however, we will not do so rigorously. In general, if an operation expects an argument of one type and does not receive one, then it will return an error object corresponding to the required type. Thus, if an operation is passed an error object as the result of a previous operation, then the error object will not be of the required type and a new error object will be returned.

5.5 Booleans

We will represent the boolean type as one:

```
def bool_type = one
```

Constructing a boolean type involves preceding a boolean value with bool_type:

```
def MAKE_BOOL = make_obj bool_type
```

which expands as:

```
λvalue.λs.(s bool_type value)
```

We can now construct the typed booleans TRUE and FALSE from the untyped versions by:

```
def TRUE = MAKE_BOOL true
```

which expands as:

```
λs.(s bool_type true)
```

and:

```
def FALSE = MAKE_BOOL false
```

which expands as:

```
λs.(s bool_type false)
```

As with the error type, the test for a boolean type involves checking for bool_type:

```
def isbool = istype bool_type
```

This definition expands as:

```
λobj.(equal (type obj) bool_type)
```

A boolean error object will be an error object with type bool_type:

```
def BOOL_ERROR = MAKE_ERROR bool_type
```

which expands as:

```
λs.(s error_type bool_type)
```

The typed function NOT should either return an error if the argument is not a boolean, or extract the value from the argument, use untyped not to complement it and make a new boolean from the result:

```
def NOT X =
 if isbool X
 then MAKE_BOOL (not (value X))
 else BOOL_ERROR
```

Similarly, the typed function AND should either return an error if either argument is non-boolean, or make a new boolean, from 'and'ing the values of the arguments:

```
def AND X Y =
 if and (isbool X) (isbool Y)
 then MAKE_BOOL (and (value X) (value Y))
 else BOOL_ERROR
```

We will now consider how these definitions bolt together by looking at:

```
AND TRUE FALSE
```

After definition replacement and initial bound variable substitution we have:

```
if and (isbool TRUE) (isbool FALSE)
then MAKE_BOOL (and (value TRUE) (value FALSE))
else BOOL_ERROR
```

First of all:

```
isbool TRUE ==

λobj.(equal (type obj) bool_type) TRUE =>

equal (type TRUE) bool_type ==

equal (λobj.(obj select_first) TRUE) bool_type ->

equal (TRUE select_first) bool_type ==

equal (λs.(s bool_type true) select_first) bool_type -> ... ->

equal bool_type bool_type => ... =>

true
```

Similarly:

```
isbool FALSE => ... =>

true
```

Thus:

```
and (isbool TRUE) (isbool FALSE) -> ... ->

and true (isbool FALSE) -> ... ->

and true true => ... =>

true
```

We now evaluate:

```
MAKE_BOOL (and (value TRUE) (value FALSE))
```

For the and:

```
value TRUE ==
λobj.(obj select_second) TRUE =>
TRUE select_second ==
λs.(s bool_type true) select_second => ... =>
true
```

and:

```
value FALSE => ... =>
false
```

so:

```
MAKE_BOOL (and (value TRUE) (value FALSE)) ->
MAKE_BOOL (and true (value FALSE)) ->
MAKE_BOOL (and true false) ->
MAKE_BOOL false ==
λvalue.λs.(s bool_type value) false =>
λs.(s bool_type false) ==
FALSE
```

5.6 Typed conditional expression

We need a typed conditional expression to handle both typed booleans and type errors in a condition:

```
def COND E1 E2 C =
 if isbool C
 then
  if value C
  then E1
  else E2
 else BOOL_ERROR
```

Note that this typed conditional function will return BOOL_ERROR if the condition is not a boolean.

We will now write:

```
IF <condition>
THEN <expression1>
ELSE <expression2>
```

instead of:

```
COND <expression1> <expression2> <condition>
```

We also need typed versions of the type testers for use with IF because iserror and isbool return the untyped true or false instead of the typed TRUE or FALSE:

```
def ISERROR E = MAKE_BOOL (iserror E)

def ISBOOL B = MAKE_BOOL (isbool B)
```

5.7 Numbers and arithmetic

We will represent the number type as two:

```
def numb_type = two
```

and a number will be a pair starting with numb_type:

```
def MAKE_NUMB = make_obj numb_type
```

MAKE_NUMB's definition expands to:

```
λvalue.λs.(s numb_type value)
```

We need an error object for arithmetic type errors:

```
def NUMB_ERROR = MAKE_ERROR numb_type
```

which expands to:

```
λs.(s error_type numb_type)
```

We also need a type tester:

```
def isnumb = istype numb_type
```

which expands to:

```
λobj.(equal (type obj) numb_type)
```

from which we can define a typed type tester:

```
def ISNUMB N = MAKE_BOOL (isnumb N)
```

Next, we can construct a typed zero:

```
def 0 = MAKE_NUMB zero
```

which expands as:

```
λs.(s numb_type zero)
```

We now need a typed successor function:

```
def SUCC N =
 if isnumb N
 then MAKE_NUMB (succ (value N))
 else NUMB_ERROR
```

to define numbers:

```
def 1 = SUCC 0

def 2 = SUCC 1

def 3 = SUCC 2
etc.
```

For example, 1 expands as:

```
SUCC 0 => ... =>

if isnumb 0
then MAKE_NUMB (succ (value N))
else NUMB_ERROR
```

First of all:

```
isnumb 0 ==

equal (type 0) numb_type ==

equal (λobj.(obj select_first) 0) numb_type =>
```

```
equal (0 select_first) numb_type ==
equal (λs.(s numb_type zero) select_first) numb_type -> ... ->
equal numb_type numb_type => ... =>
true
```

Thus, we next evaluate:

```
MAKE_NUMB (succ (value 0)) ==
MAKE_NUMB (succ (λobj.(obj select_second) 0)) ->
MAKE_NUMB (succ (0 select_second)) ==
MAKE_NUMB (succ (λs.(s numb_type zero)
                  select_second) -> ... ->
MAKE_NUMB (succ zero) ==
MAKE_NUMB one ==
λvalue.λs.(s numb_type value) one =>
λs.(s numb_type one)
```

In general, a typed number is a pair with the untyped equivalent as value.

We can now redefine the predecessor function to return an error for zero:

```
def PRED N =
 if isnumb N
 then
  if iszero (value N)
  then NUMB_ERROR
  else MAKE_NUMB ((value N) select_second)
 else NUMB_ERROR
```

Note that we return NUMB_ERROR for a non-number argument and a zero number argument. We could construct more elaborate error objects to distinguish such cases but we will not pursue this further here.

We will need a typed test for zero:

```
def ISZERO N =
 if isnumb N
 then MAKE_BOOL (iszero (value N))
 else NUMB_ERROR
```

Now we can redefine the binary arithmetic operations. They all need to test that both arguments are numbers so we will introduce an auxiliary function to do this:

```
def both_numbs X Y = and (isnumb X) (isnumb Y)
```

So, for addition, based on our earlier definition:

```
def + X Y =
 if both_numbs X Y
 then MAKE_NUMB (add (value X) (value Y))
 else NUMB_ERROR
```

and multiplication:

```
def * X Y =
 if both_numbs X Y
 then MAKE_NUMB (mult (value X) (value Y))
 else NUMB_ERROR
```

and division to take account of a zero divisor:

```
def / X Y =
 if both_numbs X Y
 then
  if iszero (value Y)
  then NUMB_ERROR
  else MAKE_NUMB (div1 (value X) (value Y))
 else NUMB_ERROR
```

and equality:

```
def EQUAL X Y =
 if both_numbs X Y
 then MAKE_BOOL (equal (value X) (value Y))
 else NUMB_ERROR
```

5.8 Characters

Let us now add characters to our types. Character values are specified exhaustively:

```
'0' is a character
'1' is a character
...
'9' is a character
```

```
'A' is a character
'B' is a character
...
'Z' is a character

'a' is a character
'b' is a character
...
'z' is a character

'.' is a character
',' is a character
...
```

It is useful to have orderings on subsequences of characters for lexicographical purposes:

```
'0' < '1'
'1' < '2'
...
'8' < '9'

'A' < 'B'
'B' < 'C'
...
'Y' < 'Z'

'a' < 'b'
'b' < 'c'
...
'y' < 'z'
```

where the ordering relation has the usual transitive property:

```
X < Z
if X < Y and Y < Z
```

It simplifies character manipulation if there is a uniform ordering overall. For example, in the ASCII character set:

```
'9' < 'A'
'Z' < 'a'
```

and most punctuation marks appear before '0' in the ordering.

We will introduce a new type for characters:

```
def char_type = four

def CHAR_ERROR = MAKE_ERROR char_type

def ischar = istype char_type

def ISCHAR C = MAKE_BOOL (ischar C)

def MAKE_CHAR = make_obj char_type
```

A character object will have type char_type. To provide the ordering, characters will be mapped onto the natural numbers so the value of a character will be an untyped number. We will use the ASCII values:

```
def '0' = MAKE_CHAR forty_eight

def '1' = MAKE_CHAR (succ (value '0'))

 ...

def '9' = MAKE_CHAR (succ (value '8'))

def 'A' = MAKE_CHAR sixty_five

def 'B' = MAKE_CHAR (succ (value 'A'))

 ...

def 'Z' = MAKE_CHAR (succ (value 'Y'))

def 'a' = MAKE_CHAR ninety_seven

def 'b' = MAKE_CHAR (succ (value 'a'))

 ...

def 'z' = MAKE_CHAR (succ (value 'y'))
```

Now we can define character ordering:

```
def CHAR_LESS C1 C2 =
 if and (ischar C1) (ischar C2)
 then MAKE_BOOL (less (value C1) (value C2))
 else CHAR_ERROR
```

and conversion from character to number:

```
def ORD C =
 if ischar C
 then MAKE_NUMB (value C)
 else CHAR_ERROR
```

and vice versa:

```
def CHAR N =
 if isnumb N
 then MAKE_CHAR (value N)
 else NUMB_ERROR
```

For example, to find the numeric equivalent of 'A':

```
ORD 'A' => ... =>
MAKE_NUMB (value 'A') ==
MAKE_NUMB (value λs.(s char_type sixty_five)) -> ... ->
MAKE_NUMB sixty_five => ... =>
λs.(numb_type sixty_five) ==
65
```

Similarly, to construct a character from the number 98:

```
CHAR 98 => ... =>
MAKE_CHAR (value 98) ==
MAKE_CHAR (value λs.(s numb_type ninety_eight)) -> ... ->
MAKE_CHAR ninety_eight => ... =>
λs.(chartype ninety_eight) ==
'b'
```

Because we have used numbers as character values, we can base character comparison on number comparison:

```
def CHAR_EQUAL C1 C2 =
 if and (ischar C1) (ischar C2)
 then MAKE_BOOL (equal (value C1) (value C2))
 else CHAR_ERROR
```

5.9 Repetitive type checking

Once we have defined typed TRUE, FALSE, ISBOOL and IF, we could define typed versions of all the other boolean operations from them, for example:

```
def NOT X =
 IF X
 THEN FALSE
 ELSE TRUE
```

```
def AND X Y =
 IF ISBOOL Y
 THEN
  IF X
  THEN Y
  ELSE FALSE
 ELSE BOOL_ERROR
```

Note that for NOT, we do not need to check explicitly that X is a boolean because the IF does so anyway. In the same way in AND, we do not need to check that X is a boolean as the second IF will perform the check.

With typed boolean operations and having defined typed 0, SUCC, PRED, ISNUMB and ISZERO, we could define the other arithmetic operations using only typed operations, for example:

```
def ADD X Y =
 IF AND (ISNUMB X) (ISNUMB Y)
 THEN ADD1 X Y
 ELSE NUMB_ERROR

rec ADD1 X Y =
 IF ISZERO Y
 THEN X
 ELSE ADD1 (SUCC X) (PRED Y)
```

Here we have defined an outer non-recursive function to check the arguments and an auxiliary recursive function to carry out the operation without argument checks. We could avoid the explicit check that Y is a number as ISZERO will do so. However, for a non-numeric argument, ISZERO (and thence the two IFs) will return a BOOL_ERROR instead of a NUMB_ERROR.

As definitions, these seem satisfactory but they would be appallingly inefficient if used as the basis of an implementation because of repetitive type checking. Consider, for example:

```
ADD 1 2
```

First of all in:

```
IF AND (ISNUMB 1) (ISNUMB 2)
```

both:

```
ISNUMB 1
```

and:

```
ISNUMB 2
```

are checked and return booleans. Next:

```
AND (ISNUMB 1) (ISNUMB 2)
```

checks that both ISNUMBs return booleans and then itself returns a boolean. Then:

```
IF AND (ISNUMB 1) (ISNUMB 2)
```

checks that AND returns a boolean.

Secondly, after ADD1 is called:

```
IF ISZERO 2
```

calls:

```
ISZERO 2
```

which checks that 2 is a number and returns a boolean. Next:

```
IF ISZERO 2
```

checks that ISZERO returns a boolean. Now, ADD1 is called recursively through:

```
ADD1 (SUCC 1) (PRED 2)
```

so:

```
IF ISZERO (PRED 2)
```

calls:

```
ISZERO (PRED 2)
```

which checks that (PRED 2) is a number and returns a boolean and then:

```
IF ISZERO (PRED 2)
```

checks that ISZERO returns a boolean. Once again, ADD1 is called recursively in:

```
ADD1 (SUCC (SUCC 1)) (PRED (PRED 2))
```

and evaluation, and type checking, continue.

Note, this example also highlights repetitive argument evaluation due to naive normal order evaluation. We will consider different approaches to argument evaluation in Chapter 8.

5.10 Static and dynamic type checking

Clearly, there is a great deal of unnecessary type checking here. Arguably, we 'know' that the function types match, so we only need to test the outer level arguments once. This is the approach we have used above in defining typed operations, where the arguments are checked before untyped operations are carried out. But how do we 'know' that the types match up? It is all very well to claim that we are using types consistently in relatively small definitions, but in developing large expressions, type mismatches will inevitably slip through, just as they do during the development of large programs. The whole point of types is to detect such inconsistencies.

As we saw above, using untyped functions is analogous to programming in a language without typed variables or type checking like machine code or BCPL. Similarly, using our fully typed functions is analogous to programming in a language where variables are untyped but objects are checked **dynamically** by each operation while a program runs, as in PROLOG and LISP.

The alternative is to introduce **static** types into the syntax of the language and then check type consistency symbolically before running programs. With symbolic checking, run-time checks are redundant. Typing may be made explicit with typed declarations, as in C and Pascal, or deduced from variable and operation use, as in ML and PS-Algol, though in the last two languages types may, and sometimes must, be specified explicitly as well.

There are well-developed theories of types that are used to define and manipulate typed objects and typed languages. Some languages provide for user defined types in a form based on such theories. For example, ML and Miranda provide for user defined types through abstract data types which, in effect, allow functional abstraction over type definitions. We will not consider these further. We will continue to define 'basic' typed functions from untyped functions and subsequently use the typed functions. What constitutes a 'basic' function will be as much a matter of expediency as theory! For pure typed functions though, the excessive type checking will remain.

5.11 Infix operators

In our notation the function always precedes the arguments in a function application. This is known as **prefix** notation and is used in LISP. Most programming languages follow traditional logic and arithmetic and allow **infix** notation as well for some binary function names. These may appear between their arguments. We will now allow the names for logical and arithmetic functions to be used as infix operators so, for example:

```
<expression1> AND <expression2> ==
 AND <expression1> <expression2>
<expression1> OR <expression2> ==
 OR <expression1> <expression2>
<expression1> + <expression2> ==
 + <expression1> <expression2>
<expression1> - <expression2> ==
 - <expression1> <expression2>
<expression1> * <expression2> ==
 * <expression1> <expression2>
<expression1> / <expression2> ==
 / <expression1> <expression2>
```

To simplify the presentation we will not introduce operator precedence or implicit associativity. Thus, strict bracketing is still required for function application arguments. For example, we write:

```
7 + (8 * 9) == + 7 (8 * 9) == + 7 (* 8 9)
```

rather than the ambiguous:

```
7 + 8 * 9
```

Some languages allow the programmer to introduce new infix binary operators with specified precedence and associativity. These include Algol 68, ML, POP-2 and PROLOG.

5.12 Case definitions and structure matching

In this chapter, we have introduced formal definitions based on the structure of the type involved. Thus, booleans are defined by listing the values TRUE and FALSE, so boolean function definitions have explicit cases for combinations of TRUE and FALSE. Numbers are defined in terms of 0 and the application of the successor function SUCC. Thus, numeric functions have base cases for 0 and recursion cases for non-zero numbers.

In general, for multi-case definitions we have written:

```
<name> <names1> = <expression1>
<name> <names2> = <expression2>
 ...
```

where <names> is a structured sequence of bound variables, constants and constructors. When a function with a multi-case definition is applied to an argument, the argument is **matched** against the structured bound variable, constant and constructor sequences, to determine which case applies. When a match succeeds for a particular case, then that case's bound variables are associated with the corresponding argument substructures for subsequent use in the case's right-hand side expression. This is known as **structure matching.**

In our functional notation, however, we have to use conditional expressions explicitly to determine the structure of an object and hence which case of a definition should be used to process it. We then use explicit selection of substructures from structured arguments. Some languages allow the direct use of case definitions and structure matching, for example PROLOG, ML and Miranda. We will extend our notation in a similar manner so a function may now take the form:

```
λ<names1>.<expression1>
 or <names2>.<expression2>
 or ...
```

and a definition simplifies to:

```
def <name> <names1> = <expression1>
 or <name> <names2> = <expression2>
 or ...
```

For recursive functions, rec is used in place of def. Note that the effect of matching depends on the order in which the cases are tried. Here, we will match cases from first to last in order. We also assume that each case is distinct from the others so at most only one match will succeed. When a case defined function is applied to an argument, if the argument matches <names1> then the result is <expression1>; if the argument matches <names2> then the result is <expression2> and so on.

For boolean functions, we will allow the use of the constants TRUE and FALSE in place of bound variables. In general, for:

```
def <name> <bound variable> =
 IF <bound variable>
 THEN <expression1>
 ELSE <expression2>
```

we will now write:

```
def <name> TRUE = <expression1>
 or <name> FALSE = <expression2>
```

Thus, negation is defined by:

```
NOT TRUE = FALSE
NOT FALSE = TRUE
```

but written as:

```
def NOT X =
 IF X
 THEN FALSE
 ELSE TRUE
```

We will now write:

```
def NOT TRUE = FALSE
 or NOT FALSE = TRUE
```

Similarly, implication is defined by:

```
IMPLIES TRUE Y = Y
IMPLIES FALSE Y = TRUE
```

but written as:

```
def IMPLIES X Y =
 IF X
 THEN Y
 ELSE TRUE
```

We will now write:

```
def IMPLIES TRUE Y = Y
 or IMPLIES FALSE Y = TRUE
```

For numbers, we will allow the use of the constant 0 and bound variables qualified by nested SUCCs in place of bound variables. In general, for:

```
rec <name> <bound variable> =
 IF ISZERO <bound variable>
 THEN <expression1>
 ELSE <expression2 using (PRED <bound variable>)>
```

we will now write:

```
rec <name> 0 = <expression1>
 or <name> (SUCC <bound variable>) =
    <expression2 using <bound variable>>
```

Thus, the sum of the first X integers is defined by:

```
SUM 0 = 0
SUM (SUCC X) = (SUCC X) + (SUM X)
```

but written as:

```
rec SUM X =
 IF ISZERO X
 THEN 0
 ELSE X + (SUM(PRED X))
```

We will now write:

```
rec SUM 0 = 0
 or SUM (SUCC X) = (SUCC X) + (SUM X))
```

Similarly, the power function is defined by:

```
POWER X 0 = 1
POWER X (SUCC Y) = X*(POWER X Y)
```

but written as:

```
rec POWER X Y =
 IF ISZERO Y
 THEN 1
 ELSE X*(POWER X (PRED Y))
```

We will now write:

```
rec POWER X 0 = 1
 or POWER X (SUCC Y) = X*(POWER X Y)
```

SUMMARY

- Types can be used to ensure that only meaningful arguments are passed to functions.
- Types can be considered informally as classes of operations and objects.
- Typed objects may be represented using type/value pairs.

- The following types have been developed:

  An error type.

  A boolean type with typed boolean operations.

  A number type with typed numeric operations.

  A character type with typed character operations.
- Typed conditional expressions with an 'IF...THEN...ELSE' notation have been introduced.
- Static and dynamic type checking can be used to avoid repetitive type checking.
- Notation has been introduced for strictly bracketed infix expressions, and for case definitions and structure matching.

Some of these topics are summarized below.

IF ... THEN ... ELSE ...

```
IF <condition>
THEN <expression1>
ELSE <expression2> ==

COND <expression1> <expression2> <condition>
```

Infix operators

```
<expression1> <operator> <expression2> ==

<operator> <expression1> <expression2>
```

Note that strict bracketing is required for nested infix expressions.

Boolean case definition

```
def <name> TRUE = <expression1>
 or <name> FALSE = <expression2> ==

def <name> <bound variable> =
 IF <bound variable>
 THEN <expression1>
 ELSE <expression2>
```

Number case definition

```
rec <name> 0 = <expression1>
 or <name> (SUCC <bound variable>) =
     <expression2 using '<bound variable>'> ==
```

```
rec <name> <bound variable> =
 IF ISZERO <bound variable>
 THEN <expression1>
 ELSE <expression2 using 'PRED <bound variable>'>
```

EXERCISES

5.1 Evaluate fully the following expressions:

(a) ISBOOL 3
(b) ISNUMB FALSE
(c) NOT 1
(d) TRUE AND 2
(e) 2 + TRUE

5.2 Signed numbers might be introduced as a new type with an extra layer of 'pairing' so that the value of a number is preceded by a boolean to indicate whether or not the number is positive or negative:

```
def signed_type = ...
def SIGN_ERROR = MAKE_ERROR signed_type
def POS = TRUE
def NEG = FALSE
def MAKE_SIGNED N SIGN = make_obj
                          signed_type
                          (make_obj SIGN N)
```

So:

```
+<number> == MAKE_SIGNED <number> POS
-<number> == MAKE_SIGNED <number> NEG
```

For example:

```
+4 == MAKE_SIGNED 4 POS
-4 == MAKE_SIGNED 4 NEG
```

Note that there are two representations for 0:

```
+0 == MAKE_SIGNED 0 POS
-0 == MAKE_SIGNED 0 NEG
```

(a) Define tester and selector functions for signed numbers:

```
def issigned N = ...       – true if N is a signed number
def ISSIGNED N = ...       – TRUE if N is a signed number
def sign N = ...           – N's sign as an untyped number
def SIGN N = ...           – N's sign as a typed number
def sign_value N = ...     – N's value as an unsigned number
def VALUE N = ...          – N's value as a signed number
def sign_iszero N = ...    – true if N is 0
```

Show that your functions work for representative positive and negative values, and 0.

(b) Define signed versions of ISZERO, SUCC and PRED:

```
def SIGN_ISZERO N = ...
def SIGN_SUCC N = ...
def SIGN_PRED N = ...
```

Show that your functions work for representative positive and negative values, and 0.

(c) Define a signed version of '+':

```
def SIGN_+ X Y = ...
```

Show that your function works for representative positive and negative values, and 0.

Chapter 6
Lists and strings

In this chapter, we are going to look at the list data structure which is used to hold variable length sequences of values. We will discuss list construction and list element access, and then use pair functions and the type representation techniques to add lists to our notation. Next, we will develop elementary functions for manipulating linear lists and simpler list notations. We will also introduce strings as lists of characters. Structure matching for list functions will be introduced and we will look at a variety of operations on linear lists. In imperative languages, arrays may be used in these applications. Finally, we will consider the use of mapping functions to generalize operations on linear lists.

6.1 Lists

Lists are general purpose data structures which are widely used within functional and logic programming. They were brought to prominence through LISP and are now found in various forms in many languages. Lists may be used in place of both arrays and record structures, for example, to build stacks, queues and tree structures.

In this chapter, we are going to introduce lists into our functional notation and look at using lists in problems where arrays might be used in other languages. In Chapter 7, we will look at using lists where record structures might be used in other languages.

Lists are variable length sequences of objects. A strict approach to lists, as in ML for example, treats them as variable length sequences of objects of the same type. We will take the relatively lax approach of LISP and PROLOG and treat them as variable length sequences of objects of mixed type. Although this is less rigorous theoretically, and makes a formal treatment more complex, it simplifies presentation and provides a more flexible structure.

Formally, a list is either empty, denoted by the unique object:

```
NIL is a list
```

or it is a constructed pair with a **head** which is any object and a **tail**, which is a list:

```
CONS H T is a list
if H is any object and T is a list
```

CONS is the traditional name for the list constructor, originally from LISP. For example, from the object 3 and the list NIL, we can construct the list:

```
CONS 3 NIL
```

with 3 in the head and NIL in the tail. From the object 2 and this list, we can construct the list:

```
CONS 2 (CONS 3 NIL)
```

with 2 in the head and the list CONS 3 NIL in the tail. From the object 1 and the previous list we can construct the list:

```
CONS  1 (CONS 2 (CONS 3 NIL))
```

with 1 in the head and the list CONS 2 (CONS 3 NIL)) in the tail, and so on.

Note that the tail of a list must be a list. Thus, all lists end, eventually, with the empty list. Note also, that the head of a list may be any object including another list, enabling the construction of nested structures, particularly trees.

If the head of a list and the heads of all of its tail lists are not lists, then the list is said to be **linear**. In LISP parlance, an object in a list which is not a list (or a function) is known as an **atom**.

The head and tail may be selected from non-empty lists:

```
HEAD (CONS H T) = H
TAIL (CONS H T) = T
```

but head and tail selection from an empty list is not permitted:

```
HEAD NIL = LIST_ERROR
TAIL NIL = LIST_ERROR
```

Consider, for example, the linear list of numbers we constructed above:

```
CONS 1 (CONS 2 (CONS 3 NIL))
```

The head of this list:

```
HEAD (CONS 1 (CONS 2 (CONS 3 NIL)))
```

is:

```
1
```

The tail of this list:

```
TAIL (CONS 1 (CONS 2 (CONS 3 NIL)))
```

is:

```
CONS 2 (CONS 3 NIL)
```

The head of the tail of the list:

```
HEAD (TAIL (CONS 1 (CONS 2 (CONS 3 NIL))))
```

is:

```
HEAD (CONS 2 (CONS 3 NIL))
```

which is:

```
2
```

The tail of the tail of the list:

```
TAIL (TAIL (CONS 1 (CONS 2 (CONS 3 NIL))))
```

is:

```
TAIL (CONS 2 (CONS 3 NIL))
```

which is:

```
CONS 3 NIL
```

The head of the tail of the tail of this list:

```
HEAD (TAIL (TAIL (CONS 1 (CONS 2 (CONS 3 NIL)))))
```

is:

```
HEAD (TAIL (CONS 2 (CONS 3 NIL)))
```

which is:

```
HEAD (CONS 3 NIL)
```

giving:

```
3
```

The tail of the tail of the tail of the list:

```
TAIL (TAIL (TAIL (CONS 1 (CONS 2 (CONS 3 NIL)))))
```

is:

```
TAIL (TAIL (CONS 2 (CONS 3 NIL)))
```

which is:

```
TAIL (CONS 3 NIL)
```

giving:

```
NIL
```

6.2 List representation

First, we define the list type:

```
def list_type = three
```

and associated tests:

```
def islist = istype list_type

def ISLIST L = MAKE_BOOL (islist L)
```

and error object:

```
def LIST_ERROR = MAKE_ERROR list_type
```

A list value will consist of a pair made from the list head and tail, so the general form of a list object with head <head> and tail <tail> will be:

```
λs.(s list_type
      λs.(s <head> <tail>))
```

Thus, we define:

```
def MAKE_LIST = make_obj list_type

def CONS H T =
 if islist T
 then MAKE_LIST λs.(s H T)
 else LIST_ERROR
```

For example:

```
CONS 1 NIL => ... =>

MAKE_LIST λs.(s 1 NIL) => ... =>

λs.(s list_type
      λs.(s 1 NIL))
```

and:

```
CONS 2 (CONS 1 NIL) => ... =>

MAKE_LIST λs.(s 2 (CONS 1 NIL)) => ... =>

λs.(s list_type
      λs.(s 2 (CONS 1 NIL))) => ... =>
```

```
λs.(s list_type
      λs.(s 2 λs.(s list_type
                    λs.(s 1 NIL))))
```

The empty list will have both head and tail set to LIST_ERROR:

```
def NIL = MAKE_LIST λs.(s LIST_ERROR LIST_ERROR)
```

so NIL is:

```
λs.(s list_type
      λs.(s LIST_ERROR LIST_ERROR))
```

Now we can use the pair selectors to extract the head and tail:

```
def HEAD L =
 if islist L
 then (value L) select_first
 else LIST_ERROR

def TAIL L =
 if islist L
 then (value L) select_second
 else LIST_ERROR
```

For example:

```
HEAD (CONS 1 (CONS 2 NIL)) ==

(value (CONS 1 (CONS 2 NIL))) select_first ==

(value λs.(s list_type
             λs.(s 1 (CONS 2 NIL)))) select_first => ... =>

λs.(s 1 (CONS 2 NIL)) select_first =>

select_first 1 (CONS 2 NIL) => ... =>

1
```

and:

```
TAIL (CONS 1 (CONS 2 NIL)) ==

(value (CONS 1 (CONS 2 NIL))) select_second ==

(value λs.(s list_type
             λs.(s 1 (CONS 2 NIL)))) select_second => ... =>

λs.(s 1 (CONS 2 NIL)) select_second =>
```

```
select_second 1 (CONS 2 NIL) => ... =>
(CONS 2 NIL)
```

and:

```
HEAD (TAIL (CONS 1 (CONS 2 NIL))) -> ... ->
HEAD (CONS 2 NIL) ==
(value (CONS 2 NIL)) select_first ==
(value λs.(list_type λs.(s 2 NIL))) select_first => ... =>
λs.(s 2 NIL) select_first => ... =>
select_first 2 NIL => ... =>
2
```

Note that both HEAD and TAIL will return LIST_ERROR from the empty list. Thus:

```
HEAD NIL ==
(value NIL) select_first => ...=>
λs.(s LIST_ERROR LIST_ERROR) select_first =>
select_first LIST_ERROR LIST_ERROR => ... =>
LIST_ERROR
```

and:

```
TAIL NIL ==
(value NIL) select_second => ...=>
λs.(s LIST_ERROR LIST_ERROR) select_second =>
select_second LIST_ERROR LIST_ERROR => ... =>
LIST_ERROR
```

The test for an empty list checks for a list with an error object in the head:

```
def isnil L =
 if islist L
 then iserror (HEAD L)
 else false
```

```
def ISNIL L =
 if islist L
 then MAKE_BOOL (iserror (HEAD L))
 else LIST_ERROR
```

6.3 Operations on lists

We will now define a variety of elementary operations on lists.

6.3.1 Linear length of a list

A list is a sequence of an arbitrary number of objects. Let us find out how many objects are in a linear list. If the list is empty, then there are 0 objects:

```
LENGTH NIL = 0
```

and otherwise there is 1 more than the number in the tail:

```
LENGTH (CONS H T) = SUCC (LENGTH T)
```

For example:

```
LENGTH (CONS 1 (CONS 2 (CONS 3 NIL))) -> ... ->
SUCC (LENGTH (CONS 2 (CONS 3 NIL))) -> ... ->
SUCC (SUCC (LENGTH (CONS 3 NIL))) -> ... ->
SUCC (SUCC (SUCC (LENGTH NIL))) -> ... ->
SUCC (SUCC (SUCC 0)) ==
3
```

In our notation, this is:

```
rec LENGTH L =
 IF ISNIL L
 THEN 0
 ELSE SUCC (LENGTH (TAIL L))
```

For example, consider:

```
LENGTH (CONS 1 (CONS 2 NIL)) -> ... ->
SUCC (LENGTH (TAIL (CONS 1 (CONS 2 NIL)))) -> ... ->
SUCC (LENGTH (CONS 2 NIL)) -> ... ->
SUCC (SUCC (LENGTH (TAIL (CONS 2 NIL)))) -> ... ->
SUCC (SUCC (LENGTH NIL)) -> ... ->
SUCC (SUCC 0) ==
2
```

Note that the selection of the tail of the list is implicit in the case notation, but must be made explicit in our current notation.

6.3.2 Appending lists

It is often useful to build one large linear list from several smaller lists. Let us append two lists so that the second is a linear continuation of the first. If the first is empty, then the result is the second:

```
APPEND NIL L = L
```

Otherwise, the result is that of constructing a new list, with the head of the first list as the head, and the result of appending the tail of the first list to the second, as the tail:

```
APPEND (CONS H T) L = CONS H (APPEND T L)
```

For example, to join:

```
CONS 1 (CONS 2 NIL)
```

to:

```
CONS 3 (CONS 4 NIL)
```

to get:

```
CONS 1 (CONS 2 (CONS 3 (CONS 4 NIL)))
```

we use:

```
APPEND (CONS 1 (CONS 2 NIL))
       (CONS 3 (CONS 4 NIL)) -> ... ->
CONS 1 (APPEND (CONS 2 NIL)
               (CONS 3 (CONS 4 NIL))) -> ... ->
```

```
CONS 1 (CONS 2 (APPEND NIL (CONS 3
                             (CONS 4 NIL)))) -> ... ->
CONS 1 (CONS 2 (CONS 3 (CONS 4 NIL)))
```

In our notation this is:

```
rec APPEND L1 L2 =
 IF ISNIL L1
 THEN L2
 ELSE CONS (HEAD L1) (APPEND (TAIL L1) L2)
```

For example, consider:

```
APPEND (CONS 1 (CONS 2 NIL)) (CONS 3 NIL) -> ... ->
CONS (HEAD (CONS 1 (CONS 2 NIL)))
     (APPEND (TAIL (CONS 1 (CONS 2 NIL)))
             (CONS 3 NIL)) -> ... ->
CONS 1 (APPEND (TAIL (CONS 1 (CONS 2 NIL)))
               (CONS 3 NIL)) -> ... ->
CONS 1 (APPEND (CONS 2 NIL) (CONS 3 NIL)) -> ... ->
CONS 1 (CONS (HEAD (CONS 2 NIL))
             (APPEND (TAIL (CONS 2 NIL))
                     (CONS 3 NIL))) -> ... ->
CONS 1 (CONS 2
              (APPEND (TAIL (CONS 2 NIL))
                      (CONS 3 NIL))) -> ... ->
CONS 1 (CONS 2 (APPEND NIL (CONS 3 NIL))) -> ... ->
CONS 1 (CONS 2 (CONS 3 NIL))
```

Note again, that the implicit list head and tail selection in the case notation is replaced by explicit selection in our current notation.

6.4 List notation

The previous examples illustrate how inconvenient it is to represent lists in a functional form: there is an excess of brackets and CONSs! In LISP, a linear list may be represented as a sequence of objects within brackets, with an implict NIL at the end, but this overloads the function application notation. LISP's simple, uniform notation for data and functions is an undoubted strength for a particular sort of programming where programs manipulate the text of other programs, but it is somewhat opaque.

We will introduce two new notations. First of all, we will follow ML and use the binary infix operator '::' in place of CONS. (LISP and Prolog use '.' as an infix concatenation operator.) Thus:

```
<expression1>::<expression2> == CONS <expression1>
                                     <expression2>
```

For example:

```
CONS 1 (CONS 2 (CONS 3 NIL)) ==

CONS 1 (CONS 2 (3::NIL)) ==

CONS 1 (2::(3::NIL)) ==

1::(2::(3::NIL))
```

Secondly, we will adopt the notation used by ML and PROLOG and represent a linear list with an implicit NIL at the end as a sequence of objects within square brackets '[' and ']', separated by commas:

```
X::NIL == [X]
X::[Y] == [X,Y]
```

For example:

```
CONS 1 NIL == 1::NIL == [1]

CONS 1 (CONS 2 NIL) == 1::(2::NIL) == 1::[2] == [1,2]

CONS 1 (CONS 2 (CONS 3 NIL)) == 1::(2::(3::NIL)) ==
1::(2::[3]) == 1::[2,3] == [1,2,3]
```

This leads naturally to:

```
NIL = [ ]
```

which is a list of no objects with an implicit NIL at the end. For example, a list of pairs:

```
CONS (CONS 5 (CONS 12 NIL))
     (CONS (CONS 10 (CONS 15 NIL))
           (CONS (CONS 15 (CONS 23 NIL))
                 (CONS (CONS 20 (CONS 45 NIL))
                       NIL)))
```

becomes the compact:

```
[[5,12],[10,15],[15,23],[20,45]]
```

As before, HEAD selects the head element:

```
HEAD (X::Y) = X
```

and TAIL selects the tail:

```
TAIL (X::Y) = Y
```

For example:

```
HEAD [[5,12],[10,15],[15,23],[20,45]] => ... =>
[5,12]
```

and:

```
TAIL [[5,12],[10,15],[15,23],[20,45]] => ... =>
[[10,15],[15,23],[20,45]]
```

and:

```
TAIL [5] => ... =>
[]
```

Note that constructing a list with CONS is not the same as using '[' and ']'. For lists constructed with '[' and ']' there is an assumed empty list at the end. Thus:

```
[<first item>,<second item>] ==
CONS <first item> (CONS <second item> NIL)
```

We may simplify long lists made from '::' by dropping intervening brackets. Thus:

```
<expression1>::(<expression2>::<expression3>) ==
<expression1>::<expression2>::<expression3>
```

For example:

```
1::(2::(3::(4::[ ]))) == 1::2::3::4::[ ] == [1,2,3,4]
```

6.5 Lists and evaluation

It is important to note that we have adopted tacitly a weaker form of β reduction with lists, because we are not evaluating fully the expressions corresponding to list representations. A list of the form:

```
<expression1>::<expression2>
```

is shorthand for:

```
CONS <expression1> <expression2>
```

which is a function application and should, strictly speaking, be evaluated. Here, we have tended to use a modified form of applicative order, where we evaluate the arguments <expression1> and <expression2> to get values <value1> and <value2>, but do not then evaluate the resulting function application:

```
CONS <value1> <value2>
```

any further. Similarly, a list of the form:

```
[<expression1>,<expression2>]
```

has been evaluated to:

```
[<value1>,<value2>]
```

but no further, even though it is equivalent to:

```
CONS <value1> (CONS <value2> NIL)
```

This should be borne in mind until we discuss evaluation in more detail in Chapter 8.

6.6 Deletion from a list

To add a new value to an unordered list, we CONS it on to the front of the list. To delete a value, we must then search the list until we find it. Thus, if the list is empty, then the value is not in it, so return the empty list:

```
DELETE X [] = []
```

It is common in list processing to return the empty list if list access fails. Otherwise, if the first value in the list is the required value, then return the rest of the list:

```
DELETE X (H::T) = T if <equal> X H
```

Otherwise, join the first value onto the result of deleting the required value from the rest of the list:

```
DELETE X (H::T) =  H::(DELETE X T) if NOT (<equal> X H)
```

Note that the comparison <equal> depends on the type of the list elements.

Suppose we are deleting from a linear list of numbers. For example, suppose we want to delete 3 from the list:

```
[1,2,3,4]
```

we use:

```
DELETE 3 [1,2,3,4] -> ... ->
1::(DELETE 3 [2,3,4]) -> ... ->
1::(2::(DELETE 3 [3,4])) -> ... ->
1::(2::[4]) -> ... ->
1::[2,4] ==
[1,2,4]
```

In our notation, the function becomes:

```
rec DELETE V L =
 IF ISNIL L
 THEN NIL
 ELSE
   IF EQUAL V (HEAD L)
   THEN TAIL L
   ELSE (HEAD L)::(DELETE V (TAIL L))
```

For example, suppose we want to delete 10 from:

```
[5,10,15,20]
```

we use:

```
DELETE 10 [5,10,15,20]
(HEAD [5,10,15,20])::(DELETE 10 (TAIL [5,10,15,20])) -> ... ->
5::(DELETE 10 ([10,15,20]))
```

```
5::(TAIL [10,15,20]) -> ... ->

5::[15,20] => ... =>

[5,15,20]
```

Note again that implicit list head and tail selection in the case notation is replaced by explicit selection in our current notation.

6.7 List comparison

Two lists are the same if they are both empty:

```
LIST_EQUAL [ ] [ ] = TRUE
LIST_EQUAL [ ] (H::T) = FALSE
LIST_EQUAL (H::T) [ ] = FALSE
```

otherwise, they are the same if the heads are equal and the tails are the same:

```
LIST_EQUAL (H1::T1) (H2::T2) = LIST_EQUAL T1 T2
                               if <equal> H1 H2

LIST_EQUAL (H1::T1) (H2::T2) = FALSE
                              if NOT (<equal> H1 H2)
```

Notice again that the comparison operation <equal> depends on the type of the list elements.

Let us compare lists of numbers, for example:

```
LIST_EQUAL [1,2,3] [1,2,3] -> ... ->

LIST_EQUAL [2,3] [2,3] -> ... ->

LIST_EQUAL [3] [3] -> ... ->

LIST_EQUAL [ ] [ ] -> ... ->

TRUE
```

In our notation, this algorithm is:

```
rec LIST_EQUAL L1 L2 =
 IF AND (ISNIL L1) (ISNIL L2)
 THEN TRUE
 ELSE
  IF OR (ISNIL L1) (ISNIL L2)
  THEN FALSE
```

```
  ELSE
   IF EQUAL (HEAD L1) (HEAD L2)
   THEN LIST_EQUAL (TAIL L1) (TAIL L2)
   ELSE FALSE
```

For example, consider:

```
LIST_EQUAL [1,2,3] [1,2,4] -> ... ->
```

This requires:

```
EQUAL (HEAD [1,2,3]) (HEAD [1,2,4])) -> ... ->
EQUAL 1 1 -> ... ->
TRUE
```

so:

```
LIST_EQUAL (TAIL [1,2,3]) (TAIL [1,2,4]) -> ... ->
LIST_EQUAL [2,3] [2,4] -> ... ->
```

This requires:

```
EQUAL (HEAD [2,3]) (HEAD [2,4])) -> ... ->
EQUAL 2 2 -> ... ->
TRUE
```

so:

```
LIST_EQUAL (TAIL [2,3]) (TAIL [2,4]) -> ... ->
LIST_EQUAL [3] [4] -> ... ->
```

This requires:

```
EQUAL (HEAD [3]) (HEAD [4])) -> ... ->
EQUAL 3 4 -> ... ->
FALSE
```

so:

```
FALSE
```

6.8 Strings

Strings are the basis of text processing in many languages. Some provide strings as a distinct type, for example, BASIC and ML. Others base strings on arrays of characters, for example, Pascal and C. Here we will introduce **strings** as linear lists of characters.

We can define a test for stringness:

```
ISSTRING [ ] = TRUE
ISSTRING (H::T) = (ISCHAR H) AND (ISSTRING T)
```

For example:

```
ISSTRING ['a','p','e'] -> ... ->

(ISCHAR 'a') AND
 (ISSTRING ['p','e']) -> ... ->

(ISCHAR 'a') AND
 ((ISCHAR 'p') AND
  (ISSTRING ['e'])) -> ... ->

(ISCHAR 'a') AND
 ((ISCHAR 'p') AND
  ((ISCHAR 'e') AND
    (ISSTRING [ ]))) -> ... ->

(ISCHAR 'a') AND
 ((ISCHAR 'p') AND
  ((ISCHAR 'e') AND
    TRUE)) -> ... ->

TRUE
```

In our notation, this function is:

```
rec ISSTRING S =
 IF ISNIL S
 THEN TRUE
 ELSE AND (ISCHAR (HEAD S)) (ISSTRING (TAIL S))
```

We will represent a string as a sequence of characters within quotes ("), for example:

```
"Here is a string!"
```

In general, the string:

```
"<character> <characters>"
```

is equivalent to the list:

```
'<character>'::"<characters>"
```

For example:

```
"cat" ==

'c'::"at" ==

'c'::'a'::"t" ==

'c'::'a'::'t'::""
```

If we represent the empty string " " as:

```
[]
```

then the string notation mirrors the compact list notation. For example:

```
"cat" ==

'c'::'a'::'t'::""==

'c'::'a'::'t'::[] ==

['c','a','t']
```

6.9 String comparison

String comparison is a type specific version of list comparison. Two strings are the same if both are empty:

```
STRING_EQUAL "" "" = TRUE
STRING_EQUAL "" (C::S) = FALSE
STRING_EQUAL (C::S) "" = FALSE
```

or if the characters in the heads are the same and the strings in the tails are the same:

```
STRING_EQUAL (C1::S1) (C2::S2) = STRING_EQUAL S1 S2
                                 if CHAR_EQUAL C1 C2

STRING_EQUAL (C1::S1) (C2::S2) = FALSE
                                 if NOT
                                  (CHAR_EQUAL C1 C2)
```

For example:

```
STRING_EQUAL "dog" "dog" ==
STRING_EQUAL ('d'::"og") ('d'::"og") -> ... ->
STRING_EQUAL "og" "og" ==
STRING_EQUAL ('o'::"g") ('o'::"g") -> ... ->
STRING_EQUAL "g" "g" ==
STRING_EQUAL ('g'::"") ('g'::"") -> ... ->
STRING_EQUAL "" "" -> ... ->
TRUE
```

In our notation:

```
rec STRING_EQUAL S1 S2 =
 IF (ISNIL S1) AND (ISNIL S2)
 THEN TRUE
 ELSE
  IF (ISNIL S1) OR (ISNIL S2)
  THEN FALSE
  ELSE
   IF CHAR_EQUAL (HEAD S1) (HEAD S2)
   THEN STRING_EQUAL (TAIL S1) (TAIL S2)
   ELSE FALSE
```

Similarly, one string comes before another if it is empty and the other is not:

```
STRING_LESS "" (C::S) = TRUE
STRING_LESS (C::S) "" = FALSE
```

or if the character in its head comes before the character in the other's head:

```
STRING_LESS (C1::S1) (C2::S2) = TRUE
                                 if CHAR_LESS C1 C2
```

or the head characters are the same and the first string's tail comes before the second string's tail:

```
STRING_LESS (C1::S1) (C2::S2) = (CHAR_EQUAL C1 C2) AND
                                 (STRING_LESS S1 S2)
                                 if NOT (CHAR_LESS C1 C2)
```

For example:

```
STRING_LESS "porridge" "potato" ==

STRING_LESS ('p'::"orridge") ('p'::"otato") -> ... ->

(CHAR_EQUAL 'p' 'p') AND
 (STRING_LESS "orridge" "otato") ==

(CHAR_EQUAL 'p' 'p') AND
 (STRING_LESS ('o'::"rridge") ('o'::"tato")) -> ... ->

(CHAR EQUAL 'p' 'p') AND
 ((CHAR_EQUAL 'o' 'o') AND
  (STRING_LESS "rridge" "tato")) ==

(CHAR EQUAL 'p' 'p') AND
 ((CHAR_EQUAL 'o' 'o') AND
  (STRING_LESS ('r'::"ridge") ('t'::"ato")) -> ... ->

(CHAR EQUAL 'p' 'p') AND
 ((CHAR_EQUAL 'o' 'o') AND
  TRUE)) -> ... ->

  TRUE
```

In our notation, this is:

```
rec STRING_LESS S1 S2 =
 IF ISNIL S1
 THEN  NOT (ISNIL S2)
 ELSE
  IF ISNIL L2
  THEN FALSE
  ELSE
   IF CHAR_LESS (HEAD S1) (HEAD S2)
   THEN TRUE
   ELSE (CHAR_EQUAL (HEAD S1) (HEAD S2)) AND
            (STRING_LESS1 (TAIL S1) (TAIL S2))
```

6.10 Numeric string to number conversion

Given a string of digits, we might wish to find the equivalent number. Note first of all, that the number equivalent to a digit is found by taking the value of '0' away from its value. For example:

```
value '0' => ... =>

forty_eight
```

so:

```
sub (value '0') (value '0') -> ... ->
sub forty_eight forty_eight => ... =>
zero
```

Similarly:

```
value '1' => ... =>
forty_nine
```

so:

```
sub (value '1') (value '0') -> ... ->
sub forty_nine forty_eight => ... =>
one
```

and:

```
value '2' => ... =>
fifty
```

so:

```
sub (value '2') (value '0') -> ... ->
sub fifty forty_eight => ... =>
two
```

and so on. Thus, we can define:

```
def digit_value d = sub (value d) (value '0')
def DIGIT_VALUE d = MAKE_NUMB (digit_value d)
```

Note also, that the value of a single digit string is the value of the digit, for example:

value of "9"

gives value of '9'

gives 9

The value of a two digit string is ten times the value of the first digit added to the value of the second digit, for example:

value of "98"

gives 10 * value of "9" + value of '8'

gives 90 + 8 = 98

The value of a three digit string is ten times the value of the first two digits added to the value of the third digit, which is ten times ten times the value of the first digit added to the second digit, added to the value of the third digit, for example:

value of "987"

gives 10 * value of "98" + value of '7'

gives 10 * (10 * value of "9" + value of '8') + value of '7'

gives 10 * (10 * 9 + 8) + 7 = 987

In general, the value of an N digit string is ten times the value of the first N−1 digits added to the value of the Nth digit. The value of an empty digit string is 0. We will implement this inside out so we can work from left to right through the string. We will keep track of the value of the first N−1 digits in another bound variable v. Each time, we will multiply v by 10 and add in the value of the Nth digit to get the value of v for processing the N+1th digit. When the string is empty we return v. To start with, v is 0. For example:

value of "987" with 0

gives value of "87" with 10 * 0 + value of '9'

gives value of "7" with 10 * (10 * 0 + value of '9') + value of '8'

gives value of " "with 10 * (10 * (10 * 0 + value of '9') +
value of '8') + value of '7'

gives 987

Thus:

```
STRING_VAL V "" = V
STRING_VAL V D::T = STRING_VAL 10*V+(DIGIT_VALUE D) T
```

For a whole string, V starts at 0, for example:

```
STRING_VAL 0 "321" ==
STRING_VAL 0 ('3'::"21") -> ... ->
STRING_VAL 10*0+(DIGIT_VALUE '3') "21" -> ... ->
STRING_VAL 3 "21" ==
STRING_VAL 3 ('2'::"1") -> ... ->
STRING_VAL 10*3+(DIGIT_VALUE '2') "1" -> ... ->
STRING_VAL 32 "1" ==
STRING_VAL 32 ('1'::"") -> ... ->
STRING_VAL 32*10+(DIGIT_VALUE '1') "" -> ... ->
STRING_VAL 321 "" -> ... ->
321
```

In our notation, using untyped arithmetic, the function is:

```
rec string_val v L =
 IF ISNIL L
 THEN v
 ELSE string_val (add (mult v ten) (digit_value (HEAD L)))
                 (TAIL L)

def STRING_VAL S = MAKE_NUMB (string_val zero S)
```

For example:

```
STRING_VAL "987" ==
MAKE_NUMB (string_val zero "987") -> ... ->
MAKE_NUMB (string_val (add
                        (mult zero ten)
                        (digit_value '9'))
                       "87") -> ... ->
MAKE_NUMB (string_val (add
                        (mult
                         (add
                          (mult zero ten)
                          (digit_value '9'))
                         ten)
                        (digit_value '8'))
```

```
                              "7") -> ... ->

MAKE_NUMB (string_val (add
                        (mult
                         (add
                          (mult
                           (add
                            (mult zero ten)
                            (digit_value '9'))
                           ten)
                          (digit_value '8'))
                          ten)
                        (digit_value '7'))
                        "") -> ... ->

MAKE_NUMB (add
           (mult
            (add
             (mult
              (add
               (mult zero ten)
               (digit_value '9'))
              ten)
             (digit_value '8'))
            ten)
           (digit_value '7'))
          -> ... ->

MAKE_NUMB (add
           (mult
            (add
             (mult nine ten)
             (digit_value '8'))
            ten)
           (digit_value '7'))
         -> ... ->

MAKE_NUMB (add
           (mult ninety_eight ten)
           (digit_value '7'))
            -> ... ->

MAKE_NUMB nine_hundred_and_eighty_seven => ... =>

987
```

6.11 Structure matching with lists

We have been defining list functions with a base case for an empty list argument, and a recursion case for a non-empty list argument, but we have translated them into explicit list selection. We will now extend our structure matching notation to lists and allow cases for the empty list NIL, and for bound variable lists built with '::' in place of bound variables. In general, for:

```
rec <name> <bound variable> =
 IF ISNIL <bound variable>
 THEN <expression1>
 ELSE <expression2 using (HEAD <bound variable>)
                      and (TAIL <bound variable>)>
```

we will write:

```
rec <name> [] = <expression1>
 or <name> (<head>::<tail>) = <expression2 using <head>
                                          and <tail>>
```

where <head> and <tail> are bound variables.

Consider the definition of the linear length of a list:

```
LENGTH [] = 0
LENGTH (H::T) = SUCC (LENGTH T)
```

which we wrote as:

```
rec LENGTH L =
 IF ISNIL L
 THEN 0
 ELSE SUCC (LENGTH (TAIL L))
```

We will now write:

```
rec LENGTH [] = 0
 or LENGTH (H::T) = SUCC (LENGTH T)
```

For example, suppose we have a list made up of arbitrarily nested lists which we want to flatten into one long linear list:

```
FLAT [[1,2,3],[[4,5],[6,7,[8,9]]]] => ... =>
[1,2,3,4,5,6,7,8,9]
```

The empty list is already flat:

```
FLAT [] = []
```

If the list is not empty, then if the head is not a list join it to the flattened tail:

```
FLAT (H::T) = H::(FLAT T) if NOT (ISLIST H)
```

Otherwise, append the flattened head to the flattened tail:

```
FLAT (H::T) = APPEND (FLAT H) (FLAT T) if ISLIST H
```

In our old notation, we would write:

```
rec FLAT L =
 IF ISNIL L
 THEN []
 ELSE
  IF NOT (ISLIST (HEAD L))
  THEN (HEAD L)::(FLAT (TAIL L))
  ELSE APPEND (FLAT (HEAD L)) (FLAT (TAIL L))
```

We will now write:

```
rec FLAT [] = []
 or FLAT (H::T) =
  IF NOT (ISLIST H)
  THEN H::(FLAT T)
  ELSE APPEND (FLAT H) (FLAT T)
```

Note that we may still need explicit conditional expressions in case definitions. For example, in FLAT above, a conditional is needed to distinguish the cases where the argument list has, or does not have, a list in its head. Structure matching can only distinguish between structural differences; it cannot distinguish between arbitrary values within structures.

Note that in LISP there are no case definitions or structure matching, so explicit list selection is necessary.

6.12 Ordered linear lists, insertion and sorting

For many applications, it is useful to hold data in some order to ease data access and presentation. Here we will look at ordered lists of data.

First of all, an ordered list is empty:

```
ORDERED [] = TRUE
```

or has a single element:

```
ORDERED [C] = TRUE
```

or has a head which comes before the head of the tail and an ordered tail:

```
ORDERED (C1::C2::L) = (<less> C1 C2) AND
                      (ORDERED (CONS C2 L))
```

For example:

```
[1,2,3]
```

is ordered because 1 comes before 2 and [2,3] is ordered because 2 comes before 3 and [3] is ordered because it has a single element.

Let us insert an item into an ordered list. If the list is empty, then the new list has the item as the sole element:

```
INSERT X [] = [X]
```

or if the item comes before the head of the list, then the new list has the item as head and the old list as tail:

```
INSERT X (H::T) = X::H::T
                  if <less> X H
```

otherwise, the new list has the head of the old list as head, and the item inserted into the tail of the old list as tail:

```
INSERT X (H::T) = H::(INSERT X T)
                  if NOT <less> X H
```

Note that this description tacitly assumes that all the items in the list are of the same type with a defined order relation <less>. For example, for lists of strings <less> will be STRING_LESS:

```
rec INSERT S [] = [S]
 or INSERT S (H::T) =
  IF STRING_LESS S H
  THEN S::H::T
  ELSE H::(INSERT S T)
```

For example:

```
INSERT "cherry" ["apple","banana","date"] => ... =>
"apple"::(INSERT "cherry" ["banana","date"]) -> ... ->
"apple"::"banana"::(INSERT "cherry" ["date"]) -> ... ->
"apple"::"banana"::"cherry"::["date"] ==
["apple","banana","cherry","date"]
```

Insertion forms the basis of a simple sort. The empty list is sorted:

```
SORT [ ] = [ ]
```

and to sort a non-empty list, insert the head into the sorted tail:

```
SORT (H::T) = INSERT H (SORT T)
```

This is a general definition but must be made type specific. Once again, we will consider a list of strings:

```
rec SORT [ ] = [ ]
 or SORT (H::T) = INSERT H (SORT T)
```

For example:

```
SORT ["cat","bat","ass"] => ... =>
INSERT "cat" (SORT ["bat","ass"]) -> ... ->
INSERT "cat" (INSERT "bat" (SORT ["ass"])) -> ... ->
INSERT "cat" (INSERT "bat" (INSERT "ass" [ ])) -> ... ->
INSERT "cat" (INSERT "bat" ["ass"]) -> ... ->
INSERT "cat" ["ass","bat"] => ... =>
["ass","bat","cat"]
```

6.13 Indexed linear list access

Arrays are often used where a linear sequence of objects is to be manipulated in terms of the linear positions of the objects in the sequence. In the same way, it is often useful to access an element of a linear list by specifying its position, relative to the start of the list. For example, in:

```
["Chris","Jean","Les","Pat","Phil"]
```

the first name is:

```
"Chris"
```

and we use:

```
HEAD ["Chris","Jean","Les","Pat","Phil"]
```

to access it, the second name is:

```
"Jean"
```

and we use:

```
HEAD (TAIL ["Chris","Jean","Les","Pat","Phil"])
```

to access it, the third name is:

```
"Les"
```

and we use:

```
HEAD (TAIL (TAIL ["Chris","Jean","Les","Pat","Phil"]))
```

to access it and so on.

In general, to access the <number>+1th name, we take the TAIL <number> times and then take the HEAD:

```
IFIND (SUCC N) (H::T) = IFIND N T
IFIND 0 (H::T) = H
```

Note that the first element is number 0, the second element is number 1 and so on, because we are basing our definition on how often the tail is taken, to make the element the head of the remaining list. If the list does not have <number> elements we will just return the empty list:

```
IFIND N [] = []
```

To summarize:

```
rec IFIND N [] = []
 or IFIND 0 (H::T) = H
 or IFIND (SUCC N) (H::T) = IFIND N T
```

For example:

```
IFIND 3 ["Chris","Jean","Les","Pat","Phil"] => ... =>

IFIND 2  ["Jean","Les","Pat","Phil"] => ... =>

IFIND 1  ["Les","Pat","Phil"] => ... =>

IFIND 0  ["Pat","Phil"] => ... =>

HEAD ["Pat","Phil"] => ... =>

"Pat"
```

Similarly, to remove a specified element from a list: if the list is empty, then return the empty list:

```
IDELETE N [] = []
```

If the specified element is at the head of the list, then return the tail of the list:

```
IDELETE 0 (H::T) = T
```

Otherwise, join the head of the list to the result of removing the element from the tail of the list, remembering that its position in the tail is one less than its position in the whole list:

```
IDELETE (SUCC N) (H::T) = H::(IDELETE N T)
```

To summarize:

```
rec IDELETE N [] = []
 or IDELETE 0 (H::T) = H
 or IDELETE (SUCC N) (H::T) = H::(IDELETE N T)
```

For example:

```
IDELETE 2 ["Chris","Jean","Les","Pat","Phil"] => ... =>

"Chris"::(IDELETE 1 ["Jean","Les","Pat","Phil"]) -> ... ->

"Chris"::"Jean":: (IDELETE 0 ["Les","Pat","Phil"]) -> ... ->

"Chris"::"Jean"::["Pat","Phil"]) ==

["Chris","Jean","Pat","Phil"]
```

New elements may be added to a list in a specified position. If the list is empty, then return the empty list:

```
IBEFORE N E [] = []
```

If the specified position is at the head of the list, then make the new element the head of a list with the old list as tail:

```
IBEFORE 0 E L = E::L
```

Otherwise, add the head of the list to the result of placing the new element in the tail of the list, remembering that the specified position is now one less than its position in the whole list:

```
IBEFORE (SUCC N) E (H::T) = H::(IBEFORE N E T)
```

To summarize:

```
rec IBEFORE N E [] = []
 or IBEFORE 0 E L = E::L
 or IBEFORE (SUCC N) E (H::T) = H::(IBEFORE N E T)
```

For example:

```
IBEFORE 2 "Jo" ["Chris","Jean","Les","Pat","Phil"] => ... =>
"Chris"::(IBEFORE 1 "Jo" ["Jean","Les","Pat","Phil"]) -> ... ->
"Chris"::"Jean":: (IBEFORE 0 "Jo" ["Les","Pat","Phil"]) -> ... ->
"Chris"::"Jean"::"Jo"::["Les","Pat","Phil"] ==
["Chris","Jean","Jo","Les","Pat","Phil"]
```

Finally, to replace the object in a specified position in a list, if the list is empty, then return the empty list:

```
IREPLACE N E [] = []
```

If the specified position is at the head, then make the replacement the head:

```
IREPLACE 0 E (H::T) = E::T
```

Otherwise, join the head of the list to the result of replacing the element in the tail, remembering that the position in the tail is now one less than the position in the whole list:

```
IREPLACE (SUCC N) E (H::T) = H::(IREPLACE N E T)
```

Note that we have not considered what happens if the list does not contain the requisite item.

To summarize:

```
rec IREPLACE N E [] = []
 or IREPLACE 0 E (H::T) = E::T
 or IREPLACE (SUCC N) E (H::T) = H::(IREPLACE N E T)
```

For example:

```
IREPLACE 2 "Jo" ["Chris","Jean","Les","Pat","Phil"] => ... =>
"Chris"::(IREPLACE 1 "Jo" ["Jean","Les","Pat","Phil"]) -> ... ->
"Chris"::"Jean"::(IREPLACE 0 "Jo" ["Les","Pat","Phil"]) -> ... ->
"Chris"::"Jean"::"Jo"::["Pat","Phil"] ==
["Chris","Jean","Jo","Pat","Phil"]
```

Alternatively, we could use DELETE to drop the old element and IBEFORE to place the new element, so:

```
IREPLACE N E L = IBEFORE N E (IDELETE N L)
```

This is much simpler but involves scanning the list twice.

6.14 Mapping functions

Many functions have similar structures. We can take advantage of this to simplify function construction, by defining abstract general purpose functions for common structures, and inserting particular functions into them to make them carry out particular processes. For example, we have defined a general purpose make_object function in Chapter 5 which we have then used to construct specialized MAKE_BOOL, MAKE_NUMB, MAKE_LIST and MAKE_CHAR functions.

For lists, such generalized functions are known as **mapping functions** because they are used to **map** a function onto the components of lists. The use of list mapping functions originated with LISP. For example, consider a function which doubles every value in a list of numbers:

```
rec DOUBLE [] = []
 or DOUBLE (H::T) = (2*H)::(DOUBLE T)
```

so:

```
DOUBLE [1,2,3] => ... =>
(2*1)::(DOUBLE [2,3]) -> ... ->
```

```
2::(2*2)::(DOUBLE [3]) -> ... ->
2::4::(2*3)::(DOUBLE [ ]) -> ... ->
2::4::6::[ ] ==
[2,4,6]
```

Now consider the function which turns all the words in a list into plurals:

```
rec PLURAL [ ] = [ ]
 or PLURAL (H::T) = (APPEND H "s")::(PLURAL T)
```

so:

```
PLURAL ["cat","dog","pig"] => ... =>
(APPEND "cat" "s")::(PLURAL ["dog","pig"]) -> ... ->
"cats"::(APPEND "dog" "s")::(PLURAL ["pig"]) -> ... ->
"cats"::"dogs"::(APPEND "pig" "s")::(PLURAL [ ]) -> ... ->
"cats"::"dogs"::"pigs"::[ ] ==
["cats","dogs","pigs"]
```

The functions DOUBLE and PLURAL both apply a function repeatedly to the consecutive heads of their list arguments. In LISP this is known as a **car** (that is, head) **mapping** because the function is mapped onto the cars of the list. We can abstract a common structure from DOUBLE and PLURAL as:

```
rec MAPCAR FUNC [ ] = [ ]
 or MAPCAR FUNC (H::T) = (FUNC H)::(MAPCAR FUNC T)
```

Thus, we can define DOUBLE as:

```
def DOUBLE = MAPCAR λX.(2*X)
```

so DOUBLE's definition expands as:

```
def DOUBLE [ ] = [ ]
 or DOUBLE (H::T) =(λX.(2*X) H)::(MAPCAR λX.(2*X) T)
```

Simplifying, we get:

```
def DOUBLE [ ] = [ ]
 or DOUBLE (H::T) = (2*H)::(MAPCAR λX.(2*X) T)
```

which is equivalent to the original:

```
rec DOUBLE [] = []
 or DOUBLE (H::T) = (2*H)::(DOUBLE T)
```

because:

```
DOUBLE == MAPCAR λX.(2*X)
```

For example:

```
DOUBLE [1,2,3] => ... =>
(λX.(2*X) 1)::(MAPCAR λX.(2*X) [2,3]) -> ... ->
2::(λX.(2*X) 2)::(MAPCAR λX.(2*X) [3]) -> ... ->
2::4::(λX.(2*X) 3)::(MAPCAR λX.(2*X) []) -> ... ->
2::4::6::[] ==
[2,4,6]
```

Similarly, we can redefine PLURAL as:

```
def PLURAL = MAPCAR λW.(APPEND W "s")
```

so expanding the definition gives:

```
def PLURAL [] = []
  or PLURAL (H::T) = (λW.(APPEND W "s") H)::
                     (MAPCAR λW.(APPEND W "s") T)
```

so, simplifying:

```
def PLURAL [] = []
  or PLURAL (H::T) = (APPEND H "s")::
                     (MAPCAR λW.(APPEND W "s") T)
```

which is equivalent to:

```
rec PLURAL [] = []
 or PLURAL (H::T) = (APPEND H "s")::(PLURAL T)
```

because:

```
PLURAL == MAPCAR λW.(APPEND W "s")
```

For example:

```
PLURAL ["cat","dog","pig"] => ... =>

(λW.(APPEND W "s") "cat")::
 (MAPCAR λW.(APPEND W "s") ["dog","pig"]) -> ... ->

"cats"::(λW.(APPEND W "s") "dog")::
 (MAPCAR λW.(APPEND W "s") ["pig"]) -> ... ->

"cats"::"dogs"::(λW.(APPEND W "s") "pig")::
 (MAPCAR λW.(APPEND W "s") [ ]) -> ... ->

 "cats"::"dogs"::"pigs"::[ ] ==

 ["cats","dogs","pigs"]
```

Consider the function which compares two equal length linear lists of strings component by component, and constructs a boolean list showing where they are the same and where they differ:

```
rec COMP [ ] [ ] = [ ]
 or COMP (H1::T1) (H2::T2) = (STRING_EQUAL H1 H2)::
                             (COMP T1 T2)
```

so:

```
COMP ["hey","diddle","diddle"] ["hey","daddle","diddle"] => ... =>

(STRING_EQUAL "hey" "hey")::
 (COMP ["diddle","diddle"] ["daddle","diddle"]) -> ... ->

TRUE::(STRING_EQUAL "diddle" "daddle")::
 (COMP ["diddle"] ["diddle"]) -> ... ->

TRUE::FALSE::(STRING_EQUAL "diddle" "diddle")::
  (COMP [ ] [ ]) -> ... ->

TRUE::FALSE::TRUE::[ ] ==

[TRUE,FALSE,TRUE]
```

Now consider the function that adds together corresponding components of two equal length linear numeric lists:

```
rec SUM2 [ ] [ ] = [ ]
 or SUM2 (H1::T1) (H2::T2) = (H1+H2)::(SUM2 T1 T2)
```

so:

```
SUM2 [1,2,3] [4,5,6] => ... =>

(1+4)::(SUM2 [2,3] [5,6]) -> ... ->
```

```
5::(2+5)::(SUM2 [3] [6]) -> ... ->

5::7::(3+6)::(SUM2 [ ] [ ]) -> ... ->

5::7::9::[ ] ==

[5,7,9]
```

The functions COMP and SUM2 both apply a function repeatedly to the consecutive heads of two list arguments to construct a new list. We can abstract a common structure from COMP and SUM2 as:

```
rec MAPCARS FUNC [ ] [ ] = [ ]
 or MAPCARS FUNC (H1::T1) (H2::T2) = (FUNC H1 H2)::
                                     (MAPCARS FUNC T1 T2)
```

Thus:

```
def COMP = MAPCARS λX.λY.(STRING_EQUAL X Y)
def SUM2 = MAPCARS λX.λY.(X+Y)
```

SUMMARY

- A list is either empty or is a constructed pair with a head and a tail.
- Lists can be represented by a list type and typed list operations.
- Elementary functions have been developed for manipulating linear lists.
- Strings are character lists with simplified notation.
- Functions have been developed for constructing ordered linear lists and for indexed list access.
- Mapping functions are used to generalize linear list operations.

Some of these topics are summarized below.

List notation

```
<expression1>::<expression2> ==
  CONS <expression1><expression2>

[<expression1>,<expression2>] ==
 <expression1>::[<expression2>]

[<expression>] == <expression>::NIL

[ ] == NIL

<expression1>::(<expression2>::<expression3>) ==

<expression1>::<expression2>::<expression3>
```

String notation

```
"<character> <characters>" == <character>::"<characters>"

"" = []
```

List case definition

```
rec <name> [] = <expression1>
 or <name> (<head>::<tail>) =
      <expression2 using '<head>' and '<tail>'> ==

rec <name> <bound variable> =
 IF ISNIL <bound variable>
 THEN <expression1>
 ELSE <expression2 using 'HEAD <bound variable>' and
                        'TAIL <bound variable>'>
```

EXERCISES

6.1 Define a concatenation function for linear lists whose elements are atoms of the same type.

6.2 (a) Write a function which indicates whether or not a list starts with a sublist. For example:

```
STARTS "The" "The cat sat on the mat." => ... =>
TRUE

STARTS "A" "The cat sat on the mat." => ... =>
FALSE
```

(b) Write a function which indicates whether or not a list contains a given sublist. For example:

```
CONTAINS "the" "The cat sat on the mat." => ... =>
TRUE

CONTAINS "the" "All cats sit on all mats." => ... =>
FALSE
```

(c) Write a function which counts how often a sublist appears in another list. For example:

```
COUNT "at" "The cat sat on the mat." => ... =>
3
```

(d) Write a function which removes a sublist from the start of a list, assuming that you know that the sublist starts the list. For example:

```
REMOVE "The " "The cat sat on the mat." => ... =>
"cat sat on the mat."
```

(e) Write a function which deletes the first occurrence of a sublist in another list. For example:

```
DELETE "sat" "The cat sat on the mat." => ... =>
"The cat  on the mat."

DELETE "lay" "The cat sat on the mat." => ... =>
"The cat sat on the mat."
```

(f) Write a function which inserts a sublist after the first occurrence of another sublist in a list. For example:

```
INSERT "sat" "cat " "The cat  on the mat." => ... =>
"The cat sat on the mat."

INSERT "sat" "fish " "The cat  on the mat." => ... =>
"The cat  on the mat."
```

(g) Write a function which replaces a sublist with another sublist in a list. For example:

```
REPLACE "sat" "lay" "The cat sat on the mat." => ... =>
"The cat lay on the mat."

REPLACE "sit" "lay" "The cat sat on the mat." => ... =>
"The cat sat on the mat."
```

6.3 (a) Write a function which merges two ordered lists to produce an ordered list. Merging the empty list with an ordered list gives that list. To merge two non-empty lists, if the head of the first comes before the head of the second, then join the head of the first onto the result of merging the tail of the first, and the second. Otherwise, join the head of the second onto the result of merging the first onto the tail of the second. For example:

```
MERGE [1,4,7,9] [2,5,8] => ... =>
[1,2,4,5,7,8,9]
```

(b) Write a function which merges a list of ordered lists. For example:

```
LMERGE [[1,4,7],[2,5,8],[3,6,9]] => ... =>
[1,2,3,4,5,6,7,8,9]
```

Chapter 7
Composite values and trees

In this chapter we are going to discuss the use of composite values to hold records of related values. Composite values will be represented as lists and composite value sequences will be processed using linear list algorithms. We will then introduce new notations to generalize list structure matching, to simplify list and composite value processing. Finally, we will look at trees and consider the use of binary tree algorithms.

7.1 Composite values

So far, we have been looking at processing sequences of single values held in lists. However, for many applications, the data is a sequence of **composite values** where each consists of a number of related subvalues. These subvalues may in turn be composite, so composite values may be nested. For example: in processing a circulation list, we need to know each person's forename and surname; in processing a stock control system, for each item in stock we need to know its name, the number in stock and the stock level at which it should be reordered; in processing a telephone directory, we need to know each person's name, address and telephone number, where the name might consist of a forename and surname.

Some languages provide special constructs for user defined composite values, for example, the Pascal RECORD, the C structure and the ML tuple. In effect, these add new types to the language. We will look at the use of ML tuples in Chapter 9.

Here, we are going to use lists to represent composite values. This is formally less rigorous than introducing a special construct, but greatly simplifies presentation.

Consider the following examples. We might represent a name consisting of a string <forename> and a string <surname> as the list:

```
[<forename>,<surname>]
```

for example:

```
["Anna","Able"]
```

or:

```
["Betty","Baker"]
```

Or, we might represent a stock item consisting of a string <item name>, an integer <stock level> and an integer <reorder level> as the list:

```
[<item name>,<stock level>,<reorder level>]
```

for example:

```
["VDU",25,10]
```

or:

```
["modem",12,15]
```

Or, we might represent a telephone directory entry consisting of a composite value name as above, a string <address> and an integer <number> as a list:

```
[[<forename>,<surname>],<address>,<number>]
```

for example:

```
[["Anna","Able"],"Accounts",1212]
```

or:

```
[["Betty","Baker"],"Boiler room",4242]
```

A sequence of composite values will be represented by a list of lists, for example, a circulation list:

```
[["Anna","Able"],
 ["Betty","Baker"],
 ["Clarice","Charlie"]]
```

or a stock list:

```
[["VDU",25,10],
 ["modem",12,15],
 ["printer",250,7]]
```

or a telephone directory:

```
[[["Anna","Able"],"Accounts",1212],
 [["Betty","Baker"],"Boiler room",4242],
 [["Clarice","Charlie"],"Customer orders",1234]]
```

7.2 Processing composite value sequences

We are using linear lists to represent composite value sequences, so we will now look at the use of linear list algorithms to process them.

Consider the following examples. Suppose that given a circulation list, we want to find someone's forename from their surname. If the list is empty, then return the empty list. If the list is not empty then if the surname matches that for the first name in the list, return the corresponding forename. Otherwise, try looking in the rest of the list.

```
rec NFIND S [ ] = [ ]
 or NFIND S (H::T) =
  IF STRING_EQUAL S (HEAD (TAIL H))
  THEN HEAD H
  ELSE NFIND S T
```

For example:

```
NFIND "Charlie" [["Anna","Able"],
                 ["Betty","Baker"],
                 ["Clarice","Charlie"]] => ... =>
NFIND "Charlie" [["Betty","Baker"],
                 ["Clarice","Charlie"]] => ... =>
NFIND "Charlie" [["Clarice","Charlie"]] => ... =>
"Clarice"
```

Or, given a stock list, suppose we want to find all the items which have the stock level below the reorder level. If the list is empty, then return the empty list. If the list is not empty, then if the first item's stock level is below the reorder level, add the first item to the result of checking the rest of the list. Otherwise, check the rest of the list:

```
rec SCHECK [ ] = [ ]
 or SCHECK (H::T) =
  IF LESS (HEAD (TAIL H)) (HEAD (TAIL (TAIL H)))
  THEN H::(SCHECK T)
  ELSE SCHECK T
```

For example:

```
SCHECK [["VDU",25,12],
        ["modem",10,12],
        ["printer",125,10],
        ["mouse",7,12]] => ... =>
SCHECK [["modem",10,12],
        ["printer",125,10],
        ["mouse",7,12]] => ... =>
["modem",10,12]::(SCHECK [["printer",125,10],
                          ["mouse",7,12]]) -> ... ->
["modem",10,12]::(SCHECK [["mouse",7,12]]) -> ... ->
["modem",10,12]::["mouse",7,12]::(SCHECK [ ]) -> ... ->
["modem",10,12]::["mouse",7,12]::[ ] ==
```

```
[["modem",10,12],
 ["mouse",7,12]]
```

Or, given a telephone directory, suppose we want to change someone's telephone number, knowing their surname. If the directory is empty, then return the empty list. If the directory is not empty, then if the required entry is the first, add a modified first entry to the rest of the entries. Otherwise, add the first entry to the result of looking for the required entry in the rest of the entries:

```
rec DCHANGE S N [] = []
 or DCHANGE S N (H::T) =
  IF STRING_EQUAL S (HEAD (TAIL (HEAD H)))
  THEN [(HEAD H),(HEAD TAIL H),N]::T
  ELSE H::(DCHANGE S N T)
```

For example:

```
DCHANGE "Charlie" 2424
 [[["Anna","Able"],"Accounts",1212],
  [["Betty","Baker"],"Boiler room",4242],
  [["Clarice","Charlie"],"Customer orders",1234]] => ... =>

[["Anna","Able"],"Accounts",1212]::
 (DCHANGE "Charlie" 2424
   [[["Betty","Baker"],"Boiler room",4242],
    [["Clarice","Charlie"],"Customer orders",1234]]) => ... =>

[["Anna","Able"],"Accounts",1212]::
 [["Betty","Baker"],"Boiler room",4242]::
  (DCHANGE "Charlie" 2424
    [[["Clarice","Charlie"],"Customer orders",1234]]) => ... =>

[["Anna","Able"],"Accounts",1212]::
 [["Betty","Baker"],"Boiler room",4242]::
  [["Clarice","Charlie"],"Customer orders",2424]::
   [] ==

[[["Anna","Able"],"Accounts",1212],
 [["Betty","Baker"],"Boiler room",4242],
 [["Clarice","Charlie"],"Customer orders",2424]]
```

7.3 Selector functions

Because composite values are being represented by lists, the above examples all depend on the nested use of the list selectors HEAD and TAIL to select subvalues from composite values. For composite values with many subvalues, this list selection becomes somewhat dense. Instead, we might define **selector** functions which are named to reflect the composite values that they operate on. These are particularly useful in LISP to simplify complex list expressions, because it lacks structure matching.

For example, for names we might define:

```
def FORENAME N = HEAD N

def SURNAME N = HEAD (TAIL N)
```

for stock items we might define:

```
def ITEM N = HEAD N

def STOCK N = HEAD (TAIL N)

def REORDER N = HEAD (TAIL (TAIL N))
```

for telephone directory entries we might define:

```
def NAME E = HEAD E

def EFORENAME E = FORENAME (NAME E)

def ESURNAME E = SURNAME (NAME E)

def ADDRESS E = HEAD (TAIL E)

def PHONE E = HEAD (TAIL (TAIL E))
```

These selector functions disguise the underlying representation and make it easier to understand the functions that use them. Consider the following examples. Given a circulation list, we might want to delete a name, knowing the surname. If the list is empty, then return the empty list. If the list is not empty, then if the surname is that of the first name, return the rest of the list. Otherwise, add the first name to the result of deleting the required name from the rest of the list:

```
rec NDELETE S [] = []
 or NDELETE S (H::T) =
  IF STRING_EQUAL S (SURNAME H)
  THEN T
  ELSE H::(NDELETE S T)
```

For example:

```
NDELETE "Charlie" [["Anna","Able"],
                   ["Betty","Baker"],
                   ["Clarice","Charlie"]] -> ... ->
```

```
["Anna","Able"]::
 (NDELETE "Charlie" [["Betty","Baker"],
                     ["Clarice","Charlie"]]) -> ... ->

["Anna","Able"]::
 ["Betty","Baker"]::
  (NDELETE "Charlie" [["Clarice","Charlie"]]) -> ... ->

["Anna","Able"]::
 ["Betty","Baker"]::
  [] ==

[["Anna","Able"],
 ["Betty","Baker"]]
```

Or, given a stock control list, we might want to increment the stock level, knowing the item name. If the list is empty, then return the empty list. If the list is not empty, then if the first item is the required one, increment its stock level and add the changed item to the rest of the list. Otherwise, add the first item to the result of searching the rest of the list:

```
rec SINCREMENT I V [] = []
 or SINCREMENT I V (H::T) =
  IF STRING_EQUAL I (ITEM H)
  THEN [(ITEM H),(STOCK H)+V,(REORDER H)]::T
  ELSE H::(SINCREMENT I V T)
```

For example:

```
SINCREMENT "modem" 10 [["VDU",25,12],
                       ["modem",10,12],
                       ["printer",125,10]] -> ... ->

["VDU",25,12]::
 (SINCREMENT "modem" 10 [["modem",10,12],
                         ["printer",125,10]]) -> ... ->

["VDU",25,12]::
 ["modem",20,12]::
  [["printer",125,10]] ==

 [["VDU",25,12],
  ["modem",20,12]
  ["printer",125,10]]
```

Or, given a telephone directory, we might want to add a new entry in alphabetical surname order. If the directory is empty, then make a new directory from the new entry. If the directory is not empty, then if the new entry comes before the first entry, add it to the front of the directory. Otherwise, add the first entry to the result of adding the new entry to the rest of the directory:

```
rec DINSERT E [] = [E]
 or DINSERT E (H::T) =
  IF STRING_LESS (ESURNAME E) (ESURNAME H)
  THEN E::H::T
  ELSE H::(DINSERT E T)
```

For example:

```
DINSERT
  [["Chris","Catnip"],"Credit",3333]
  [[["Anna","Able"],"Accounts",1212],
   [["Betty","Baker"],"Boiler room",4242],
   [["Clarice","Charlie"],"Customer orders",2424]] => ... =>
[["Anna","Able"],"Accounts",1212]::
 (DINSERT
    [["Chris","Catnip"],"Credit",3333]
    [[["Betty","Baker"],"Boiler room",4242],
     [["Clarice","Charlie"],"Customer orders",2424]]) -> ... ->
[["Anna","Able"],"Accounts",1212]::
 [["Betty","Baker"],"Boiler room",4242]::
 (DINSERT
    [["Chris","Catnip"],"Credit",3333]
    [[["Clarice","Charlie"],"Customer orders",2424]]) -> ... ->
[["Anna","Able"],"Accounts",1212]::
 [["Betty","Baker"],"Boiler room",4242]::
  [["Chris","Catnip"],"Credit",3333]::
   [[["Clarice","Charlie"],"Customer orders",2424]] ==
[[["Anna","Able"],"Accounts",1212],
 [["Betty","Baker"],"Boiler room",4242],
 [["Chris","Catnip"],"Credit",3333],
 [["Clarice","Charlie"],"Customer orders",2424]]
```

7.4 Generalized structure matching

In Chapter 6, we introduced structure matching into function definitions. Objects are defined in terms of constant base cases and structured recursion cases. Thus, function definitions have base cases with constants, instead of bound variables, for matching against constant arguments, and recursion cases with structured bound variables for matching against structured arguments. In particular, for list processing, we have used bound variable lists of the form:

```
[]
```

for matching against the empty list, and of the form:

```
(H::T)
```

so that H matches the head of a list argument and T matches the tail. We will now allow arbitrary bound variable lists for matching against arbitrary list arguments. The bound variable lists may contain implicit or explicit empty lists for matching against empty lists in arguments. Consider the following examples. We can use structure matching to redefine the circulation list selector functions:

```
def FORENAME [F,S] = F
def SURNAME [F,S] = S
```

Here, the bound variable list:

```
[F,S] == F::S::NIL
```

matches the argument list:

```
[<forename>,<surname>] == <forename>::<surname>::NIL
```

so:

```
F == <forename>
S == <surname>
```

We can also pick up the forename and surname from the first entry in a list of names, by structure matching with the bound variable list:

```
([F,S]::T)
```

so [F,S] matches the head of the list and T matches the tail. For example, we might count how often a given forename occurs in a circulation list. If the list is empty then the count is 0. If the list is not empty, then if the forename matches that for the first entry, add 1 to the count for the rest of the list. Otherwise, return the count for the rest of the list:

```
rec NCOUNT N [] = 0
 or NCOUNT N ([F,S]::T) =
  IF STRING_EQUAL N F
  THEN 1 + (NCOUNT N T)
  ELSE (NCOUNT N T)
```

Or, we can redefine the stock control selector functions as:

```
def ITEM [I,S,R] = I
def STOCK [I,S,R] = S
def REORDER [I,S,R] = R
```

Here, the bound variable list:

```
[I,S,R] == I::S::R::NIL
```

matches the argument:

```
[<item name>,<stock level>,<reorder level>] ==
 <item name>::<stock level>::<reorder level>::NIL
```

so:

```
I == <item name>
S == <stock level>
R == <reorder level>
```

We can use the bound variable list:

```
([I,S,R]::T)
```

to match against a stock control list so that [I,S,R] matches the first item and T matches the rest of the list. For example, we might find all the items that need to be reordered. If the list is empty, then return the empty list. If the list is not empty, and if the first item needs to be reordered, then add it to those to be reordered in the rest of the list. Otherwise, return those to be reordered in the rest of the list:

```
rec REORD [] = 0
 or REORD ([I,S,R]::T) =
  IF LESS S R
```

```
   THEN [I,S,R]::(REORD T)
   ELSE REORD T
```

We can redefine the telephone directory selector functions as:

```
def NAME [N,A,P] = N

def EFORENAME [[F,S],A,P] = F

def ESURNAME [[F,S],A,P] = S

def ADDRESS [N,A,P] = A

def PHONE [N,A,P] = P
```

Here, the bound variable list:

```
[N,A,P] == N::A::P::NIL
```

matches the argument list:

```
[<name>,<address>,<number>] ==
 <name>::<address>::<number>::NIL
```

so:

```
N == <name>
A == <address>
P == <number>
```

Similarly, the bound variable list:

```
[[F,S],A,P] == (F::S::NIL)::A::P::NIL
```

matches the argument list:

```
[[<forename>,<surname>],<address>,<phone>] ==
 (<forename>::<surname>::NIL)::<address>::<phone>::NIL
```

so:

```
F == <forename>
S == <surname>
```

A bound variable list of the form:

```
[N,A,P]::T
```

can be used to match a directory so [N,A,P] matches the first entry and T matches the rest of the directory. For example, we might sort the directory in telephone number order using the insertion sort from Chapter 6:

```
rec DINSERT R [ ] = [R]
 or DINSERT [N1,A1,P1] ([N2,A2,P2]::T) =
  IF LESS P1 P2
  THEN [N1,A1,P1]::[N2,A2,P2]::T
  ELSE [N2,A2,P2]::(DINSERT [N1,A1,P1] T)
rec DSORT [ ] = [ ]
 or DSORT (H::T) = DINSERT H (DSORT T)
```

7.5 Local definitions

It is often useful to introduce new name/value associations for use within an expression. Such associations are said to be **local** to the expression and are introduced by **local definitions.** Two forms of local definition are used in functional programming and they are both equivalent to function application.

Consider:

```
λ<name>.<body> <argument>
```

This requires the replacement of all free occurrences of <name> in <body> with <argument> before <body> is evaluated. Alternatively, <name> and <argument> might be thought of as being associated throughout the evaluation of <body>. This might be written in a bottom-up style as:

```
let <name> = <argument>
in <body>
```

or in a top-down style as:

```
<body>
where <name> = <argument>
```

We will use the bottom-up let form of local definition, on the grounds that things should be defined before they are used.

7.6 Matching composite value results

The use of bound variable lists greatly simplifies the construction of functions which return composite value results represented as lists. For example, suppose we have a list of forename/surname pairs and we wish to split it into separate lists of forenames and surnames. We could scan the list to pick up the forenames and again to pick up the surnames. Alternatively, we can pick them both up at once.

To split an empty list, return an empty forename list and an empty surname list. Otherwise, split the tail and put the forename from the head pair onto the forename list from the tail, and the surname from the head pair onto the surname list from the tail:

```
rec SPLIT [] = []::[]
 or SPLIT ([F,S]::L) =
  let (FLIST::SLIST) = SPLIT L
  in  ((F::FLIST)::(S::SLIST))
```

Note that at each stage, SPLIT is called recursively on the tail to return a list of lists. This is then separated into the lists FLIST and SLIST, the items from the head pair are added and a new list of lists is returned. For example:

```
SPLIT [["Allan","Ape"],["Betty","Bat"],["Colin","Cat"]] => ... =>
let (FLIST::SLIST) = SPLIT [["Betty","Bat"],["Colin","Cat"]]
in (("Allan"::FLIST)::("Ape"::SLIST))
```

The first recursive call to SPLIT involves:

```
SPLIT [["Betty","Bat"],["Colin","Cat"]] => ... =>
let (FLIST::SLIST) = SPLIT [["Colin","Cat"]]
in (("Betty"::FLIST)::("Bat"::SLIST))
```

The second recursive call to SPLIT involves:

```
SPLIT [["Colin","Cat"]] => ... =>
let (FLIST::SLIST) = SPLIT []
in (("Colin"::FLIST)::("Cat"::SLIST))
```

The third recursive call to SPLIT is the last:

```
SPLIT [] => ... =>
[]::[]
```

so the recursive calls start to return:

```
      let (FLIST::SLIST) = [ ]::[ ]
      in (("Colin"::FLIST)::("Cat"::SLIST)) => ... =>
      (("Colin"::[ ])::("Cat"::[ ])) ==
      (["Colin"]::["Cat"]) => ... =>
   let (FLIST::SLIST) = (["Colin"]::["Cat"])
   in (("Betty"::FLIST)::("Bat"::SLIST)) => ... =>
   (("Betty"::["Colin"])::("Bat"::["Cat"])) ==
   (["Betty","Colin"]::["Bat","Cat"]) => ... =>
let (FLIST::SLIST) = (["Betty","Colin"]::["Bat","Cat"])
in (("Allan"::FLIST)::("Ape"::SLIST)) => ... =>
(("Allan"::["Betty","Colin"])::("Ape"::["Bat","Cat")]) ==
(["Allan","Betty","Colin"]::["Ape","Bat","Cat"])
```

We can simplify this further by making the local variables, FLIST and SLIST, additional bound variables:

```
rec SPLIT [ ] L = L
 or SPLIT ([F,S]::L) (FLIST::SLIST) = SPLIT L ((F::FLIST)::
                                               (S::SLIST))
```

On the recursive call, the variables FLIST and SLIST will pick up the lists (F::FLIST) and (S::SLIST) from the previous call. Initially, FLIST and LIST are both empty. For example:

```
SPLIT [["Diane","Duck"],
       ["Eric","Eagle"],["Fran","Fox"]] ([ ]::[ ]) => ... =>
SPLIT [["Eric","Eagle"],["Fran","Fox"]] (["Diane"]::["Duck"]) => ... =>
SPLIT [["Fran","Fox"]] (["Eric","Diane"]::["Eagle","Duck"]) => ... =>
SPLIT [ ] (["Fran","Eric","Diane"]::["Fox","Eagle","Duck"]) => ... =>
(["Fran","Eric","Diane"]::["Fox","Eagle","Duck"])
```

Note that we have picked up the list components in reverse order, because we have added the heads of the argument lists into the new lists, before processing the tails of the argument lists.

The bound variables FLIST and SLIST are known as **accumulation variables** because they are used to accumulate partial results.

7.7 List inefficiency

Linear lists correspond well to problems involving flat sequences of data items but are relatively inefficient to access and manipulate. This is because accessing an item always involves skipping past the preceding items. In long lists this becomes extremely time consuming. For a list with N items, if we assume that each item is just as likely to be accessed as any other item, then:

> to access the 1st item, skip 0 items;
> to access the 2nd item, skip 1 item;
> ...
> to access the N − 1th item, skip N − 2 items;
> to access the Nth item, skip N − 1 items

Thus, on average it is necessary to skip:

$$(1+\ldots(N-2)+(N-1)) / N = (N * N / 2) / N = N / 2$$

items. For example, to find one item in a list of 1000 items, it is necessary to skip 500 items on average.

Sorting using the insertion technique above is far worse. For a worst case sort with N items in complete reverse order:

> to place the 1st item, skip 0 items;
> to place the 2nd item, skip 1 item;
> ...
> to place the N − 1th item, skip N − 2 items;
> to place the Nth item, skip N − 1 items

Thus, in total it is necessary to skip:

$$1+\ldots(N-2)+(N-1) = N * N / 2$$

items. For example, for a worst case sort of 1000 items, it is necessary to skip 500 000 items.

Remember, for searching and sorting in a linear list, each skip involves a comparison between a list item and a required or new item. If the items are strings, then comparison is character by character, so the number of comparisons can be rather big for relatively short lists.

Note that we have been considering naive linear list algorithms. For particular problems, if there is a known ordering on a sequence of values, then it may be possible to represent the sequence as an ordered list of ordered subsequences. For example, a sequence of strings might be represented as list of ordered sublists, with a sublist for each letter of the alphabet.

7.8 Trees

Trees are general purpose nested structures which enable far faster access to ordered sequences than linear lists. Here, we are going to look at how trees may be modelled using lists. To begin with, we will introduce the standard tree terminology.

A **tree** is a nested data structure consisting of a hierarchy of **nodes**. Each node holds one data item and has **branches** to **subtrees** which are in turn composed of nodes. The first node in a tree is called the **root**. A node with empty branches is called a **leaf**. Often, a tree has the same number of branches in each node. If there are N branches then the tree is said to be **N-ary**.

If there is an ordering relationship on the tree, then each subtree consists of nodes whose items have a common relationship to the original node's item. Note that ordering implies that the node items are all the same type. Ordered linear sequences can be held in a tree structure which enables far faster access and update.

We are now going to look specifically at **binary** trees. A binary tree node has two branches, called the **left** and **right** branches, to binary subtrees. Formally, the empty tree, which we will denote as EMPTY, is a binary tree:

```
EMPTY is a binary tree
```

and a tree consisting of a node with an item and two subtrees is a binary tree if the subtrees are binary trees:

```
NODE ITEM L R is a binary tree
if L is a binary tree and R is a binary tree
```

We will model a binary tree using lists. We will represent EMPTY as NIL:

```
def EMPTY = NIL
def ISEMPTY = ISNIL
```

and a node as a list of the item and the left and right branches:

```
def NODE ITEM L R = [ITEM,L,R]
```

The item and the subtrees may be selected from nodes:

```
ITEM (NODE I L R) = I
LEFT (NODE I L R) = L
RIGHT (NODE I L R) = R
```

but no selection may be made from empty trees:

```
ITEM EMPTY = TREE_ERROR
LEFT EMPTY = TREE_ERROR
RIGHT EMPTY = TREE_ERROR
```

Note that we cannot use these equations directly as functions, as we have not introduced trees as a new type into our notation. Instead, we will model tree functions with list functions using LIST_ERROR for TREE_ERROR:

```
def TREE_ERROR = LIST_ERROR

def ITEM EMPTY = TREE_ERROR
 or ITEM [I,L,R] = I

def LEFT EMPTY = TREE_ERROR
 or LEFT [I,L,R] = L

def RIGHT EMPTY = TREE_ERROR
 or RIGHT [I,L,R] = R
```

Note that we can use EMPTY in structure matching because it is the same as NIL.

7.9 Adding values to ordered binary trees

In an ordered binary tree, the left subtree contains nodes whose items come before the original node's item in some ordering, and the right subtree contains nodes whose items come after the original node's item in that ordering. Each subtree is itself an ordered binary tree.

Thus, to add an item to an ordered binary tree, if the tree is empty, then make a new node with empty branches for the item:

```
TADD I EMPTY = NODE I EMPTY EMPTY
```

If the item comes before the root node item, then add it to the left subtree:

```
TADD I (NODE NI L R) = NODE NI (TADD I L) R
                       if <less> I NI
```

Otherwise, add it to the right subtree:

```
TADD I (NODE NI L R) = NODE NI L (TADD I L)
                       if NOT (<less> I NI)
```

Thus, for a binary tree of integers:

```
rec TADD I EMPTY =  NODE I EMPTY EMPTY
 or TADD I [NI,L,R] =
  IF LESS I NI
  THEN NODE NI (TADD I L) R
  ELSE NODE NITEM L (TADD I R)
```

For example, to add 7 to an empty tree:

```
TADD 7 EMPTY
```

The tree is empty, so a new node is constructed:

```
[7,EMPTY,EMPTY]
```

To add 4 to this tree:

```
TADD 4 [7,EMPTY,EMPTY]
```

4 comes before 7, so it is added to the left subtree:

```
[7,(TADD 4 EMPTY),EMPTY] -> ... ->

[7,
 [4,EMPTY,EMPTY],
 EMPTY
]
```

To add 9 to this tree:

```
TADD 9 [7,
 [4,EMPTY,EMPTY],
 EMPTY
]
```

9 comes after 7, so it is added to the right subtree:

```
[7,
 [4,EMPTY,EMPTY],
 (TADD 9 EMPTY)
] -> ... ->

[7,
 [4,EMPTY,EMPTY],
 [9,EMPTY,EMPTY]
]
```

To add 3 to this tree:

```
TADD 3 [7,
        [4,EMPTY,EMPTY],
        [9,EMPTY,EMPTY]
       ]
```

3 comes before 7, so it is added to the left subtree:

```
[7,
 (TADD 3 [4,EMPTY,EMPTY]),
 [9,EMPTY,EMPTY]
]
```

3 comes before 4, so it is added to the left subtree:

```
[7,
 [4,
  (TADD 3 EMPTY),
  EMPTY
 ],
 [9,EMPTY,EMPTY]
] -> ... ->
[7,
 [4,
  [3,EMPTY,EMPTY],
  EMPTY
 ],
 [9,EMPTY,EMPTY]
]
```

To add 5 to this tree:

```
TADD 5 [7,
        [4,
         [3,EMPTY,EMPTY],
         EMPTY
        ],
        [9,EMPTY,EMPTY]
       ]
```

5 comes before 7, so it is added to the left subtree:

```
[7,
 (TADD 5 [4,
          [3,EMPTY,EMPTY],
          EMPTY
         ]),
 [9,EMPTY,EMPTY]
]
```

Now, 5 comes after 4, so it is added to the right subtree:

```
[7,
 [4,
  [3,EMPTY,EMPTY],
  (TADD 5 EMPTY)
 ],
 [9,EMPTY,EMPTY]
] -> ... ->

[7,
 [4,
  [3,EMPTY,EMPTY],
  [5,EMPTY,EMPTY]
 ],
 [9,EMPTY,EMPTY]
]
```

To add an arbitrary list of numbers to a tree, if the list is empty, then return the tree. Otherwise, add the tail of the list to the result of adding the head of the list to the tree:

```
rec TADDLIST [ ] TREE = TREE
 or TADDLIST (H::T) TREE = TADDLIST T (TADD H TREE)
```

Thus:

```
TADDLIST [7,4,9,3,5,11,6,8] EMPTY -> ... ->

[7,
 [4,
  [3,EMPTY,EMPTY],
  [5,
   EMPTY,
   [6,EMPTY,EMPTY]
  ]
 ],
 [9,
```

```
  [8,EMPTY,EMPTY],
  [11,EMPTY,EMPTY]
 ]
]
```

7.10 Binary tree traversal

Having added values to an ordered tree, it may be useful to extract them in some order. This involves **walking** or **traversing** the tree picking up the node values. From our definition of an ordered binary tree, all the values in the left subtree for a node are less than the node value, and all the values in the right subtree for a node are greater than the node value. Thus, to extract the values in ascending order, we need to traverse the left subtree, pick up the node value and then traverse the right subtree:

```
TRAVERSE (NODE I L R) =
 APPEND (TRAVERSE L) (I::(TRAVERSE R))
```

Traversing an empty tree returns an empty list:

```
TRAVERSE EMPTY = []
```

Using lists instead of a tree type:

```
rec TRAVERSE EMPTY = []
 or TRAVERSE [I,L,R] =
    APPEND (TRAVERSE L) (I::(TRAVERSE R))
```

We will illustrate this with an example. To ease presentation, we may evaluate several applications at the same time at each stage:

```
TRAVERSE [7,
          [4,
           [3,EMPTY,EMPTY],
           [5,EMPTY,EMPTY]
          ],
          [9,EMPTY,EMPTY]
         ] -> ... ->

APPEND (TRAVERSE [4,
                  [3,EMPTY,EMPTY],
                  [5,EMPTY,EMPTY]
                 ])
       (7::(TRAVERSE [9,EMPTY,EMPTY])) -> ... ->
```

```
APPEND (APPEND (TRAVERSE [3,EMPTY,EMPTY])
                (4::(TRAVERSE [5,EMPTY,EMPTY])))
       (7::(APPEND (TRAVERSE EMPTY)
                   (9::(TRAVERSE EMPTY)))) -> ... ->

APPEND (APPEND (APPEND (TRAVERSE EMPTY)
                       (3::(TRAVERSE EMPTY)))
               (4::(APPEND (TRAVERSE EMPTY)
                           (5::(TRAVERSE EMPTY))))
       (7::(APPEND (TRAVERSE EMPTY)
                   (9::(TRAVERSE EMPTY)))) -> ... ->

APPEND (APPEND (APPEND [ ]
                       (3::[ ]))
               (4::(APPEND [ ]
                           (5::[ ])))
       (7::(APPEND [ ]
                   (9::[ ]))) -> ... ->

APPEND (APPEND [3]
               (4::[5]))
       (7::[9]) -> ... ->

APPEND [3,4,5] [7,9] -> ... ->

[3,4,5,7,9]
```

7.11 Binary tree search

Once a binary tree has been constructed, it may be searched to find out whether or not it contains a value. The search algorithm is very similar to the addition algorithm above. If the tree is empty, then the search fails:

```
TFIND V EMPTY = FALSE
```

If the tree is not empty, and the required value is the node value, then the search succeeds:

```
TFIND V (NODE NV L R) = TRUE if <equal> V NV
```

Otherwise, if the required value comes before the node value, then try the left branch:

```
TFIND V (NODE NV L R) = TFIND V L if <less> V NV
```

if not, try the right branch:

```
TFIND V (NODE NV L R) = TFIND V R if NOT (<less> V NV)
```

For a binary integer tree:

```
rec TFIND V EMPTY = ""
 or TFIND V [NV,L,R] =
  IF EQUAL V NV
  THEN TRUE
  ELSE
    IF LESS V NV
    THEN TFIND V L
    ELSE TFIND V R
```

For example:

```
TFIND 5 [7,
         [4,
          [3,EMPTY,EMPTY],
          [5,EMPTY,EMPTY]
         ],
         [9,EMPTY,EMPTY]
        ] -> ... ->
TFIND 5 [4,
         [3,EMPTY,EMPTY],
         [5,EMPTY,EMPTY]
        ] -> ... ->
TFIND 5 [5,EMPTY,EMPTY] -> ... ->
TRUE
```

And as a further example:

```
TFIND 2 [7,
         [4,
          [3,EMPTY,EMPTY],
          [5,EMPTY,EMPTY]
         ],
         [9,EMPTY,EMPTY]
        ] -> ... ->
TFIND 2 [4,
         [3,EMPTY,EMPTY],
         [5,EMPTY,EMPTY]
        ] -> ... ->
```

```
TFIND 2 [3,EMPTY,EMPTY] -> ... ->
TFIND 2 EMPTY -> ... ->
FALSE
```

7.12 Binary trees of composite values

Binary trees, like linear lists, may be used to represent ordered sequences of composite values. Each node holds one composite value from the sequence and the ordering is determined by one subvalue. The tree addition functions above may be modified to work with composite values. For example, we might hold the circulation list of names in a binary tree in surname order. Adding a new name to the tree involves comparing the new surname with the node surnames:

```
rec CTADD N EMPTY = [N,EMPTY,EMPTY]
 or CTADD [F,S] [[NF,NS],L,R] =
  IF STRING_LESS S NS
  THEN [[NF,NS],(CTADD [F,S] L),R]
  ELSE [[NF,NS],L,(CTADD [F,S] R)]

rec CTADDLIST [ ] TREE = TREE
 or CTADDLIST (H::T) TREE = CTADDLIST T (CTADD H TREE)
```

For example:

```
CTADDLIST
  [["Mark","Monkey"],
   ["Graham","Goat"],
   ["Quentin","Quail"],
   ["James","Jaguar"],
   ["David","Duck]] EMPTY -> ... ->

[["Mark","Monkey"],
  [["Graham","Goat"],
   [["David","Duck"],EMPTY,EMPTY],
   [["James","Jaguar"],EMPTY,EMPTY]
  ],
  [["Quentin","Quail"],EMPTY,EMPTY]
]
```

The tree traversal function above may be applied to binary trees with arbitrary node values as it only inspects branches during traversal. For example:

```
TRAVERSE [["Mark","Monkey"],
          [["Graham","Goat"],
           [["David","Duck"],EMPTY,EMPTY],
           [["James","Jaguar"],EMPTY,EMPTY]
          ],
          [["Quentin","Quail"],EMPTY,EMPTY]
         ]-> ... ->

[["David","Duck"],
 ["Graham","Goat"],
 ["James","Jaguar"],
 ["Mark","Monkey"],
 ["Quentin","Quail"]]
```

Finally, the tree search function above may be modified to return some or all of a required composite value. For example, we might find the forename corresponding to a surname, using the surname to identify the required node:

```
rec CTFIND S EMPTY = ""
 or CTFIND S [[NF,NS],L,R] =
  IF STRING_EQUAL S NS
  THEN NF
  ELSE
   IF STRING_LESS S NS
   THEN CTFIND S L
   ELSE CTFIND S R
```

For example:

```
CTFIND "Duck" [["Mark","Monkey"],
               [["Graham","Goat"],
                [["David","Duck"],EMPTY,EMPTY],
                [["James","Jaguar"],EMPTY,EMPTY]
               ],
               [["Quentin","Quail"],EMPTY,EMPTY]
              ] -> ... ->

CTFIND "Duck" [["Graham","Goat"],
                 [["David","Duck"],EMPTY,EMPTY],
                 [["James","Jaguar"],EMPTY,EMPTY]
              ] -> ... ->

CTFIND "Duck" [["David","Duck"],EMPTY,EMPTY]

"David"
```

7.13 Binary tree efficiency

When an ordered binary tree is formed from a value sequence, each node holds an ordered subsequence. Every subnode on the left branch of a node contains values which are less than the node's value, and every subnode on the right branch contains values which are greater than the node's value. Thus, when searching a tree for a node, given a value, the selection of one branch discounts all the subnodes, and hence all the values, on the other branch. The number of comparisons required to find a node depends on how many layers of subnodes there are between the root and that node.

A binary tree is said to be **balanced** if for any node, the number of values in both branches is the same. For a balanced binary tree, if a node holds N values then there are (N−1)/2 values in its left branch and (N−1)/2 in the its right branch. Thus, the total number of branch layers depends on how often the number of values can be halved. This suggests that in general, if:

$$2^L <= N < 2^{L+1}$$

then:

$$N \text{ values} == \log_2(N)+1 == L+1 \text{ layers}$$

For example:

```
 1 value  == 1 node                   == 1 layer
                                      == log2( 1)+1
 3 values == 1 node + 2 *  1 value    == 2 layers
                                      == log2( 3)+1
 7 values == 1 node + 2 *  3 values   == 3 layers
                                      == log2( 7)+1
15 values == 1 node + 2 *  7 values   == 4 layers
                                      == log2(15)+1
31 values == 1 node + 2 * 15 values   == 5 layers
                                      == log2(31)+1
63 values == 1 node + 2 * 31 values   == 6 layers
                                      == log2(63)+1
...
```

For a balanced tree of 1000 items it is necessary to go down 10 layers, making 10 comparisons, in the worst case.

Note that we have considered perfectly balanced trees. However, the algorithms discussed above do not try to maintain balance and so the 'shape' of a tree depends on the order in which values are added. In general, trees built with our simple algorithm will not be balanced. Indeed, in the worst case, the algorithm builds a linear list, ironically, when the values are already in order:

```
TADDLIST [4,3,2,1] -> ... ->

[1,
 EMPTY,
 [2,
  EMPTY,
  [3,
   EMPTY,
   [4,EMPTY,EMPTY]
  ]
 ]
]
```

We will not consider the construction of balanced trees here.

7.14 Curried and uncurried functions

In imperative languages, like Pascal and C, we use procedures and functions declared with several formal parameters and we cannot separate a procedure or function from the name with which it is declared. Here, however, all our functions are built from nested λ functions with single bound variables: names and definitions of name/function associations are just a convenient simplification.

Our notation for function definitions and applications has led us to treat a name associated with a nested function as if it were a function with several bound variables. Now we have introduced another form of multiple bound variables through bound variable lists. In fact, nested functions of single bound variables and functions with multiple bound variables are equivalent. The technique of defining multi-parameter functions as nested single parameter functions was popularized by the American mathematician Haskell Curry and nested single parameter functions are called **curried** functions.

We can construct functions to transform a curried function into an uncurried function and vice versa. For a function f with a bound variable list containing two bound variables:

```
def curry f x y = f [x,y]
```

will convert from uncurried form to curried form. For example, consider:

```
def SUM_SQ1 [X,Y] = (X*X)+(Y*Y)
```

Then for:

```
def curry_SUM_SQ = curry SUM_SQ1
```

the right-hand side expands as:

```
λf.λx.λy.(f [x,y]) SUM_SQ1 =>
λx.λy.(SUM_SQ1 [x,y])
```

so:

```
def curry_SUM_SQ x y = SUM_SQ1 [x,y]
```

The use of curry_SUM_SQ with nested arguments is the same as the use of SUM_SQ1 with an argument list.

Similarly, for a function g with a single bound variable a, which returns a function with single bound variable b:

```
def uncurry g [a,b] = g a b
```

will convert from curried form to uncurried form. For example, with:

```
def SUM_SQ2 X Y = (X*X)+(Y*Y)
```

then for:

```
def uncurry_SUM_SQ = uncurry SUM_SQ2
```

the right-hand side expands as:

```
λg.λ[a,b].(g a b) SUM_SQ2 =>
λ[a,b].(SUM_SQ2 a b)
```

so:

```
def uncurry_SUM_SQ [a,b] = SUM_SQ2 a b
```

The use of uncurry_SUM_SQ with an argument list is equivalent to the use of SUM_SQ2 with nested arguments.

The functions curry and uncurry are inverses. For an arbitrary function:

```
<function>
```

consider:

```
uncurry (curry <function>) ==
λg.λ[a,b].(g a b) (λf.λx.λy.(f [x,y]) <function>) ->
λg.λ[a,b].(g a b) λx.λy.(<function> [x,y]) =>
λ[a,b].(λx.λy.(<function> [x,y]) a b)
```

which simplifies to:

```
λ[a,b].(<function> [a,b]) ==
<function>
```

Here we have used a form of η reduction to simplify:

```
λ[<name1>,<name2>]).(<expression> [<name1>,<name2>])
```

to:

```
<expression>
```

Similarly:

```
curry (uncurry <function>) ==
λf.λx.λy.(f [x,y]) (λg.λ[a,b].(g a b) <function>) ->
λf.λx.λy.(f [x,y]) λ[a,b].(<function> a b) =>
λx.λy.(λ[a,b].(<function> a b) [x,y])
```

which simplifies to:

```
λx.λy.(<function> x y) ==
<function>
```

Again we have used a form of η reduction to simplify:

```
λ<name1>.λ<name2>.(<expression> <name1> <name2>)
```

to:

```
<expression>
```

7.15 Partial application

We have been using a technique which is known as **partial application** where a multi-parameter function is used to construct another multi-parameter function by providing arguments for only some of the parameters. We have taken this for granted because we use nested single bound variable functions. For example, in Chapter 5 we defined the function:

```
def istype t obj = equal t (type obj)
```

to test an object's type against an arbitrary type t, and then constructed:

```
def isbool = istype bool_type

def isnumb = istype numb_type

def ischar = istype char_type

def islist = istype list_type
```

to test whether an object was a boolean, number, character or list by 'filling in' the bound variable t with an appropriate argument. This creates no problems for us, because istype is a function of one bound variable which returns a function of one bound variable. However, many imperative languages with multi-parameter procedures and functions do not usually allow procedures and functions as objects in their own right (POP-2 and PS-Algol are exceptions), and so partial application is not directly available. However, an equivalent form is based on defining a new function or procedure with less parameters, which calls the original procedure or function with some of its parameters 'filled in'. For example, had istype been defined in Pascal as:

```
FUNCTION ISTYPE(T:TYPEVAL,OBJ:OBJECT):BOOLEAN
BEGIN
      ISTYPE := (T = TYPE(OBJ))
END
```

assuming that Pascal allowed the construction of appropriate types, then ISBOOL might be defined by:

```
FUNCTION ISBOOL(O:OBJECT):BOOLEAN
BEGIN
      ISBOOL := ISTYPE(BOOL_TYPE,O)
END
```

and ISNUMB by:

```
FUNCTION ISNUMB(OBJ:OBJECT):BOOLEAN
BEGIN
      ISNUMB := ISTYPE(NUMB_TYPE,OBJ)
END
```

and so on. In our notation, it is as if we had defined:

```
def istype (t::o)= equal t (type o)
def isbool obj = istype (bool_type::obj)
def isnumb obj = istype (numb_type::obj)
```

and so on. Here, we have explicitly provided a value for the second bound variable o from the new single bound variable obj.

7.16 Structures, values and functions

We have now developed a general purpose, fairly high level functional programming notation which has many similarities to real functional languages, as we shall see when we look at LISP and Standard ML in Chapters 9 and 10. However, it is important to remember that our notation is based on adding syntactic layers to λ calculus and that we can always translate functions in the fairly high level notation back to λ calculus by simple-minded substitution.

We appear to distinguish functions and values and structures but everything is ultimately pure λ functions which we interpret in different ways. Thus, our notation contains no syntactic or type checking controls over the combination of functions and values and structures. In particular, we cannot constrain the use of selector and constructor functions for data structures to appropriate arguments. The weak type checking introduced in Chapter 5 will detect many clashes when functions are evaluated, but we can, in principle, still apply anything to anything else. The result will be a λ expression but it may not have a plausible interpretation.

SUMMARY

- Composite values may be represented as lists.
- Selector functions are used to simplify composite value manipulation.
- List structure matching has been generalized.
- Notations for local definitions have been introduced.
- The efficiency of naive linear list algorithms has been shown to be limited.
- Trees may be represented as nested lists.
- Functions have been developed to manipulate ordered binary trees.

- The efficiency of balanced binary trees has been shown to be better than that of linear lists.
- Curried and uncurried functions and partial application have been considered.

Some of these topics are summarized below.

Generalized structure matching

```
def <name> [<name1>,<name2>,<name3> ... ] =
        <expression using '<name1>', '<name2>',
                          '<name3>' ... > ==
def <name> <bound variable> =
        <expression using 'HEAD <bound variable>',
                          'HEAD (TAIL <bound variable>)',
                          'HEAD
                            (TAIL (TAIL <bound variable>))' ... >
```

Local definitions

```
let <name> = <expression1>
in <expression2> ==
<expression2>
where <name> = <expression1> ==
λ<name>.<expression2> <expression1>
```

Curried and uncurried functions

```
λ<name1>.λ<name2>...λ<nameN>.<body> ==
λ[<name1>,<name2> ... <nameN>].<body>
```

EXERCISES

7.1 The time of day might be represented as a list with three integer fields for hours, minutes and seconds:

```
[<hours>,<minutes>,<seconds>]
```

For example:

```
[17,35,42] == 17 hours 35 minutes 42 seconds
```

Note that:

```
24 hours == 0 hours
1 hour == 60 minutes
1 minute == 60 seconds
```

(a) Write functions to convert from a time of day to seconds and from seconds to a time of day. For example:

```
TOO_SECS [2,30,25] => ... => 9025

FROM_SECS 48975 => ... => [13,36,15]
```

(b) Write a function which increments the time of day by one second. For example:

```
TICK [15,27,18] => ... => [15,27,19]
TICK [15,44,59] => ... => [15,45,0]
TICK [15,59,59] => ... => [16,0,0]
TICK [23,59,59] => ... => [0,0,0]
```

(c) In a shop, each transaction at a cash register is time stamped. Given a list of transaction details, where each is a string followed by a time of day, write a function which sorts them into ascending time order. For example:

```
TSORT [["coffee",[12,19,57]],
       ["bread",[18,22,48]],
       ["orange juice",[10,12,35]],
       ["bananas",[15,47,19]]] => ... =>

[["orange juice",[10,12,35]],
 ["coffee",[12,19,57]],
 ["bananas",[15,47,19]]
 ["bread",[18,22,48]]]
```

7.2 (a) Write a function which compares two integer binary trees.

(b) Write a function which indicates whether or not one integer binary tree contains another as a subtree.

(c) Write a function which traverses a binary tree to produce a list of node values in descending order.

7.3 Strictly bracketed integer arithmetic expressions:

```
<expression> ::= (<expression> + <expression>) |
                 (<expression> + <expression>) |
```

(<expression> + <expression>) |
(<expression> + <expression>) |
<number>

might be represented by a nested list structure so:

```
(<expression1> + <expression2>) ==
 [<expression1>,"+",<expression2>]
(<expression1> - <expression2>) ==
 [<expression1>,"-",<expression2>]
(<expression1> * <expression2>) ==
 [<expression1>,"*",<expression2>]
(<expression1> / <expression2>) ==
 [<expression1>,"/",<expression2>]
 <number> == <number>
```

For example:

```
3 == 3
(3 * 4) == [3,"*",4]
((3 * 4) - 5) == [[3,"*",4],"-",5]
((3 * 4) - (5 + 6)) == [[3,"*",4],"-",[5,"+",6]]
```

Write a function which evaluates a nested list representation of an arithmetic expression. For example:

```
EVAL 3 => ... => 3
EVAL [3,"*",4] => ... => 12
EVAL [[3,"*",4],"-",5] => ... => 7
EVAL [[3,"*",4],"-",[5,"+",6]] => ... => 11
```

Chapter 8
Evaluation

In this chapter, we are going to look at evaluation order in more detail. We will consider the relative merits of applicative and normal order evaluation and see that applicative order is generally more efficient than normal order. We will also see that applicative order evaluation may lead to non-terminating evaluation sequences where normal order evaluation terminates, with our representations of conditional expressions and recursion. We will then note that the halting problem is undecidable, so it is not possible to tell whether or not the evaluation of an arbitrary λ expression terminates. We will also survey the Church-Rosser theorems which show that normal and applicative order evaluation order are equivalent but that normal order is more likely to terminate. Finally, we will look at lazy evaluation which combines the best features of normal and applicative orders.

8.1 Termination and normal form

A λ expression which cannot be reduced any further is said to be in **normal form.** Our definition of β reduction in Chapter 2 implied that evaluation of an expression terminates when it is no longer a function application. This will not reduce an expression to normal form. Technically, we should go on evaluating the function body until it contains no more function applications. Otherwise, expressions which actually reduce to the same normal form appear to be different. For example, consider:

```
λx.x λa.(a a)
```

and:

```
λf.λa.(f a) λs.(s s)
```

The first reduces as:

```
λx.x λa.(a a) =>
λa.(a a)
```

and the second as:

```
λf.λa.(f a) λs.(s s) =>
λa.(λs.(s s) a)
```

Evaluation using our definition of β reduction terminates with two different final forms. If, however, we continue to evaluate the body of the second:

```
λa.(λs.(s s) a) =>
λa.(a a)
```

we can see that they are actually identical.

To be more formal, a reducible function application expression is called a **redex.** An expression is in normal form when it contains no more redexes.

In λ calculus theory, there are two named intermediate normal forms which are related to our approach. For **weak head normal form**, the expression is a function whose body may be a redex. Thus, the body may be a function application which may be reduced further. For **head normal form**, the expression may be a function but with a body which is not a redex. The body may still be a function application:

```
(<function expression> <argument expression>)
```

but it may not be reduced further because its <function expression> is not reducible to a function. Note that in head normal form, the <argument expression> may be further reducible.

We will still tend to stop evaluation when we reach a recognizable function or function application. Thus, for lists we will continue to leave:

```
<value1>::<value2>
```

as it is, instead of translating to the function application:

```
CONS <value1> <value2>
```

and continuing with evaluation.

8.2 Normal order

Normal order β reduction requires the evaluation of the leftmost redex in an expression. For a function application, this will evaluate the function expression and then carry out the substitution of the unevaluated argument. Normal order evaluation has the effect of delaying the evaluation of applications, which are in turn arguments for other applications, and may result in the multiple evaluation of expressions. Consider, for example:

```
rec ADD X Y =
  IF ISZERO Y
  THEN X
  ELSE ADD (SUCC X) (PRED Y)
```

Now, if we evaluate:

```
ADD 1 (ADD 1 2) => ... =>

IF ISZERO (ADD 1 2)
THEN 1
ELSE ADD (SUCC 1) (PRED (ADD 1 2)) => ... =>

ADD (SUCC 1) (PRED (ADD 1 2)) => ... =>

IF ISZERO (PRED (ADD 1 2))
THEN SUCC 1
ELSE ADD (SUCC (SUCC 1))
         (PRED (PRED (ADD 1 2))) => ... =>

ADD (SUCC (SUCC 1)) (PRED (PRED (ADD 1 2))) => ... =>

IF ISZERO (PRED (PRED (ADD 1 2)))
THEN SUCC (SUCC 1)
```

```
ELSE ADD (SUCC (SUCC (SUCC 1)))
         (PRED (PRED (PRED (ADD 1 2)))) => ... =>

ADD (SUCC (SUCC (SUCC 1)))
    (PRED (PRED (PRED (ADD 1 2)))) => ... =>

IF ISZERO (PRED (PRED (PRED (ADD 1 2))))
THEN SUCC (SUCC (SUCC 1))
ELSE ... => ... =>

SUCC (SUCC (SUCC 1)) ==

4
```

is returned. ADD 1 2 has been evaluated four times even though it only appeared once originally. In the initial call to ADD, the argument ADD 1 2 is not evaluated because it is not a leftmost application. Instead, it completely replaces the bound variable Y throughout ADD's body. Thereafter, it is evaluated in the IF condition, but when ADD is called recursively it becomes an argument for the call, so evaluation is again delayed. We also had to evaluate other applications repeatedly, for example PRED (ADD 1 2) in the condition, although we did not highlight these.

In general, for normal order evaluation, an unevaluated application argument will replace all occurrences of the associated bound variable in the function body. Each replacement is the site of initiation of potential additional evaluation. Clearly, the more often the bound variable appears in the body, the more argument evaluation may multiply.

8.3 Applicative order

Applicative order β reduction of an application requires the evaluation of both the function and the argument expressions. More formally, this involves the evaluation of the leftmost redex free of internal redexes. For a function application, this will result in each argument only being evaluated once. For example, evaluating our previous example in applicative order:

```
ADD 1 (ADD 1 2) ->

ADD 1 3 -> ... ->

IF ISZERO 3
THEN 1
ELSE ADD (SUCC 1) (PRED 3) -> ... ->

ADD (SUCC 1) (PRED 3) ->

ADD 2 2 ->
```

```
IF ISZERO 2
THEN 2
ELSE ADD (SUCC 2) (PRED 2) -> ... ->

ADD (SUCC 2) (PRED 2) ->

ADD 3 1 -> ... ->

IF ISZERO 1
THEN 3
ELSE ADD (SUCC 3) (PRED 1) -> ... ->

ADD (SUCC 3) (PRED 1) ->

ADD 4 0 -> ... ->

IF ISZERO 0
THEN 4
ELSE ... -> ... ->

4
```

Here, argument evaluation appears to be minimized. For example, ADD 1 2 and PRED (ADD 1 2) are only evaluated once.

8.4 Consistent applicative order use

We have actually been using applicative order somewhat selectively; in particular, we are still evaluating IFs in normal order. Let us look at the previous example again, this time using untyped arithmetic to simplify things:

```
rec add x y =
 if iszero y
 then x
 else add (succ x) (pred y)
```

Now, consider:

```
succ (add one two)
```

First of all, the argument add one two is evaluated:

```
add one two  -> ... ->

if iszero two
then one
else add (succ one) (pred two)
```

Now, recall our definition of if as a syntactic simplification of:

```
def cond e1 e2 c = c e1 e2
```

Thus, after substitution for add's bound variables, add 1 2 becomes:

```
cond one (add (succ one) (pred two)) (iszero two) -> ... ->

cond one (add two one) (iszero two)
```

so now we have to evaluate:

```
add two one -> ... ->

if iszero one
then two
else add (succ two) (pred one)
```

Again, replacing if with cond we have:

```
cond two (add (succ two) (pred one)) (iszero one) -> ... ->

cond two (add three zero) (iszero one)
```

so we have to evaluate:

```
add three zero -> ... ->

if iszero zero
then three
else add (succ three) (pred zero)
```

which translates to:

```
cond three (add (succ three) (pred zero)) (iszero zero) -> ... ->

cond three (add four zero) (iszero zero)
```

so we have to evaluate:

```
add four zero
```

and so on. Evaluation will never terminate! This is not because pred zero is defined to be zero. The evaluation would still not terminate even if an error were returned from pred zero. Rather, the problem lies with consistent applicative order use.

if is just another function. When an if is used it is the same as calling a function in an application: all arguments must be evaluated before substitution takes place. Recursive functions are built of ifs, so if an if's argument is itself a recursive function call, then in applicative order, argument evaluation will recur indefinitely. This does not occur with normal order evaluation because the recursive function call argument to if is not evaluated until it is selected in the if's body.

We can see the same difficulty more starkly with our construction for recursion. Recall:

```
def recursive f = λs.(f (s s)) λs.(f (s s))
```

and consider, for an arbitrary function <function>:

```
recursive <function> ==

λf.(λs.(f (s s)) λs.(f (s s))) <function> ->

λs.(<function> (s s)) λs.(<function> (s s)) ->

<function>
 (λs.(<function> (s s)) λs.(<function> (s s))) ->

<function>
 (<function>
  (λs.(<function> (s s)) λs.(<function> (s s)))) ->

<function>
 (<function>
  (<function>
   (λs. (<function> (s s)) λs.(<function> (s s)))))
```

and so on. Again this will not terminate. This does not arise with normal order because argument evaluation is delayed. In this example, the self-application will depend on <function> calling itself recursively and is delayed until the recursive call is encountered as the leftmost application.

8.5 Delaying evaluation

With consistent applicative order use, we need to find some means of delaying argument evaluation explicitly. One way, as we saw when we discussed recursion, is to add an extra layer of abstraction to make an argument into a function body and then extract the argument with explicit function application to evaluate the body. For example, for an if, we need to delay evaluation of the then and else options until one is selected. We might try changing an if to take the form:

```
def cond e1 e2 c = c λdummy.e1 λdummy.e2
```

to delay the evaluation of e1 and e2. Here, the idea is that if the condition c is true then:

```
λdummy.e1
```

is selected and if the condition is false then:

```
λdummy.e2
```

is selected. Unfortunately, this will not work. Remember:

```
def cond e1 e2 c = c λdummy.e1 λdummy.e2
```

is shorthand for:

```
def cond = λe1.λe2.λc.(c λdummy.e1 λdummy.e2)
```

so when cond is called, the arguments corresponding to e1 and e2 are evaluated before being passed to:

```
λdummy.e1
```

and:

```
λdummy.e2
```

Instead, we have to change our notation for if. Now:

```
if <condition>
then <true choice>
else <false choice>
```

will be replaced by:

```
cond λdummy.<true choice> λdummy.<false choice> <condition>
```

Thus <true choice> and <false choice> will be inserted straight into the call to cond without evaluation. We have introduced the delay through a textual substitution technique which is equivalent to normal order evaluation.

Alternatively, we could redefine our def notation so it behaves like a **macro** definition and then our first attempt would work. A macro is a text substitution function. When a macro is called, occurrences of its bound variables in its body are replaced by the arguments. Macros are used for abstraction in programming languages, for example in C and in many assembly languages, but result in the addition of in-line text rather than layers of procedure or functions calls. Here:

```
def <name> <bound variables> = <body>
```

might introduce the macro <name>, and subsequent occurrences of:

```
<name> <arguments>
```

would require the replacement of <bound variables> in <body> with the corresponding <arguments> followed by the evaluation of the resulting <body>. This would introduce macro expansion in a form equivalent to normal order evaluation.

Here, we will redefine if ... then ... else .... Now, true and false must be changed to force the evaluation of the selected option. This suggests:

```
def true x y = x identity
def false x y = y identity
```

For arbitrary expressions <expression1> and <expression2>:

```
if true
then <expression1>
else <expression2> ==
cond λdummy.<expression1> λdummy.<expression2>
                                        true -> ... ->
true λdummy.<expression1> λdummy.<expression2> -> ... ->
λdummy.<expression1> identity ->
<expression1>
```

and:

```
if false
then <expression1>
else <expression2> ==
cond λdummy.<expression1> λdummy.<expression2>
                                        false -> ... ->
false λdummy.<expression1> λdummy.<expression2> -> ... ->
λdummy.<expression2> identity ->
<expression2>
```

This delayed evaluation approach is similar to the use of **thunks** to implement Algol 60 call-by-name. A thunk is a parameterless procedure which is produced for a call-by-name parameter, to delay evaluation of that parameter until the parameter is encountered in the procedure or function body.

Note that to follow this through, all definitions involving booleans also have to change to accommodate the new forms for true and false. However, we will not consider this further. We could also use abstraction to build an applicative order version of recursive: we will not consider this further either.

The effect of using abstraction to delay evaluation is to reintroduce the multiple evaluation associated with normal order. Any expression which is delayed by abstraction must be evaluated explicitly. Thus, if bound variable substitution places a delayed expression in several places then it must be explicitly evaluated in each place. Clearly, for functional language implementations based on applicative order evaluation some compromises must be made. For example, LISP is usually implemented with applicative order evaluation but the conditional operator cond implicitly delays evaluation of its arguments. Thus, the definition of recursive functions causes no problems. LISP also provides the quote and eval operators to delay and force expression evaluation explicitly.

8.6 Evaluation termination, the halting problem, evaluation equivalence and the Church-Rosser theorems

We have tacitly assumed that there is some equivalence between normal and applicative order and we have switched between them indiscriminately. There are, however, differences. We have seen that normal order may lead to repetitive argument evaluation and that applicative order may not terminate. Of course, normal order may not terminate either. One of our first examples:

```
λs.(s s) λs.(s s)
```

showed us this.

In general, there is no way of telling whether or not the evaluation of an expression will ever terminate. This was shown originally by Alan Turing who devised a formal model for computing based on what are known as Turing machines. Turing proved that it is impossible to construct a Turing machine to tell whether or not an arbitrary Turing machine halts: in formal terminology, the halting problem for Turing machines is undecidable.

Church's thesis hypothesized that all descriptions of computing are equivalent. Thus, any result for one applies to the others as well. In particular, it has been shown that λ calculus and Turing machines are equivalent: for every Turing machine there is an equivalent λ expression and vice versa. Thus, the undecidability of the halting problem applies to λ calculus as well, so there is no way to tell if evaluation of an arbitrary λ expression terminates. In principle, we can go on evaluating individual λ expressions in the hope that evaluation will terminate, but there is no way of being sure that it will.

To return to normal and applicative order reduction: two theorems by Church and Rosser show that they are interchangeable but that normal order gives a better guarantee of evaluation termination.

The first Church-Rosser theorem shows that every expression has a unique normal form. Thus, if an expression is reduced using two different evaluation orders and both reductions terminate, then they both lead to the same normal form. For example, if normal and applicative order reductions of an expression both terminate, then they produce the same final result. This suggests that we can use normal and applicative orders as we choose.

The second Church-Rosser theorem shows that if an expression has a normal form, then it may be reached by normal order evaluation. In other words, if any evaluation order will terminate then normal order evaluation is guaranteed to terminate.

These theorems suggest that normal order evaluation is the better choice if finding the normal form for an expression is an overriding consideration.

As we have seen, there are practical advantages in the selective use of applicative order and in not evaluating function bodies, even though the first reduces the likelihood of termination, and the second may stop evaluation before a normal form is reached. We will discuss evaluation strategies for real functional languages in subsequent chapters.

8.7 Infinite objects

The evaluation delay with normal order evaluation enables the construction of infinite structures. For example, for lists with normal order evaluation, if a CONS constructs a list from a recursive call, then evaluation of that call is delayed until the corresponding field is selected. To illustrate this, we will use typeless versions of CONS, HEAD and TAIL:

```
def cons h t s = s h t

def head l = l λx.λy.x

def tail l = l λx.λy.y
```

Now, let us define the list of all numbers:

```
rec numblist n = cons n (numblist (succ n))
def numbers = numblist zero
```

In normal order, numbers' definition leads to:

```
numblist zero => ... =>
cons zero (numblist (succ zero)) => ... =>
λs.(s zero (numblist (succ zero)))
```

Now:

```
head numbers => ... =>
λs.(s zero (numblist (succ zero))) λx.λy.x => ... =>
zero
```

and:

```
tail numbers => ... =>
λs.(s zero (numblist (succ zero))) λx.λy.y => ... =>
numblist (succ zero) => ... =>
λs.(s (succ zero) (numblist (succ (succ zero))))
```

so:

```
head (tail numbers) => ... =>
(tail numbers) λx.λy.x => ... =>
λs.(s (succ zero) (numblist (succ (succ zero)))) λx.λy.x => ... =>
(succ zero)
```

and:

```
tail (tail numbers) => ... =>
(tail numbers) λx.λy.y => ... =>
λs.(s (succ zero) (numblist (succ (succ zero)))) λx.λy.y => ... =>
numblist (succ (succ zero)) => ... =>
λs.(s (succ (succ zero)) (numblist (succ (succ (succ zero)))))
```

In applicative order, this definition would not terminate because the call to numblist would recur indefinitely. In normal order though, we have the multiple evaluation of succ zero, succ (succ zero) and so on. In addition, the list is recalculated up to the required value every time a value is selected from it.

8.8 Lazy evaluation

Lazy evaluation is a method of delaying expression evaluation which avoids multiple evaluation of the same expression. Thus, it combines the advantages of normal order and applicative order evaluation. With lazy evaluation, an expression is evaluated when its value is needed, that is, when it appears in the function position in a function application. However, after evaluation, all copies of that expression are updated with the new value.

Lazy evaluation requires some means of keeping track of multiple copies of expressions. We will give each bound pair in an expression a unique subscript. During expression evaluation, when a bound pair replaces a bound variable it retains its subscript, but the bound pair containing the variable and copies of it, and the surrounding bound pairs and their copies, are given consistent new subscripts. For example, consider:

$(\lambda s.(s\ s)_1\ (\lambda x.x\ \lambda y.y)_2)_3$

To evaluate the outer bound pair 3, the argument bound pair 2 is copied into the function body bound pair 1 which is renumbered 4:

$((\lambda x.x\ \lambda y.y)_2\ (\lambda x.x\ \lambda y.y)_2)_4$

Note that bound pair 2 occurs twice in bound pair 4. To evaluate bound pair 4, first evaluate the function expression which is bound pair 2:

$(\lambda x.x\ \lambda y.y)_2$ =>

$\lambda y.y$

and then replace all occurrences of it in bound pair 4 with the new value and renumber bound pair 4 as 5 to get:

$(\lambda y.y\ \lambda y.y)_5$

Finally, evaluate bound pair 5:

```
λy.y
```

Note that bound pair 2:

```
(λx.x λy.y)2
```

has only been evaluated once even though it occurs in two places.

We can now see a substantial saving in normal order evaluation with recursion. To simplify presentation, we will only number bound pairs which may be evaluated several times and we will not expand everything into λ functions. We will also use applicative order to simplify things occasionally.

Consider addition once again:

```
rec ADD X Y =
 IF ISZERO Y
 THEN X
 ELSE ADD (SUCC X) (PRED Y)
```

For the evaluation of:

```
ADD 2 2 => ... =>

IF ISZERO 2
THEN 2
ELSE ADD (SUCC 2) (PRED 2)1 => ... =>

ADD (SUCC 2) (PRED 2)1 => ... =>

IF ISZERO (PRED 2)1
THEN (SUCC 1)
ELSE ADD (SUCC (SUCC 2)) (PRED (PRED 2)1)
```

evaluate:

```
ISZERO (PRED 2)1
```

which leads to the evaluation of bound pair 1:

```
(PRED 2)1 => ... => 1
```

which replaces all other occurrences of bound pair 1:

```
IF ISZERO 1
THEN SUCC 1
ELSE ADD (SUCC (SUCC 2)) (PRED 1)2 => ... =>
```

ADD (SUCC (SUCC 2)) $(\text{PRED } 1)_2$ => ... =>

IF ISZERO $(\text{PRED } 1)_2$
THEN (SUCC (SUCC 2))
ELSE ADD (SUCC (SUCC (SUCC 2))) (PRED $(\text{PRED } 1)_2$)

so:

ISZERO $(\text{PRED } 1)_2$

is evaluated which involves the evaluation of bound pair 2:

$(\text{PRED } 1)_2$ => ... => 0

which replaces all other occurrences of bound pair 2:

IF ISZERO 0
THEN SUCC (SUCC 2)
ELSE ADD (SUCC (SUCC (SUCC 2))) (PRED 0) => ... =>

SUCC (SUCC 2) => ... =>

0

Here, the evaluation of arguments is delayed, as for normal order, but an argument is only evaluated once, as for applicative order.

Lazy evaluation avoids repetitive evaluation in infinite lists, as the list is extended whenever head or tail evaluation occurs. This makes them useful when the same values are going to be used repeatedly and one wants to avoid recalculating them every time. For example, we might define a function to calculate squares as:

```
def SQ X = X * X
```

Every time we required the square of a number, SQ would calculate it afresh. We could put the squares into an infinite list:

```
rec SQLIST N = (SQ N)::(SQLIST (SUCC N))

def SQUARES = SQLIST 0
```

so SQUARES is:

(SQ 0)::(SQLIST $(\text{SUCC } 0)_1)_2$

Here, we have only labelled the recursive extension of the list.

We now construct functions to select values from the list:

```
rec IFIND N L =
 IF ISZERO N
 THEN HEAD L
 ELSE IFIND (PRED N) (TAIL L)
def SQUARE N = IFIND N SQUARES
```

Now, if a particular square has been selected before, then it is already in the list and is therefore selected. Otherwise, selection forces evaluation until the required value is found. This forced evaluation then leaves new values in the extended list ready for another value selection. For example, consider:

```
SQUARE 2 => ...=>
IFIND 2 SQUARES => ... =>
IFIND 1 (TAIL SQUARES) => ... =>
IFIND 0 (TAIL (TAIL SQUARES)) => ... =>
HEAD (TAIL (TAIL SQUARES))
```

Now, the inner selection of:

```
TAIL SQUARES ==
TAIL ((SQ 0)::(SQLIST (SUCC 0)_1)_2)
```

results in the the selection of bound pair 2:

```
(SQLIST (SUCC 0)_1)_2
```

The next level of selection:

```
TAIL (TAIL SQUARES) => ... =>
TAIL (SQLIST (SUCC 0)_1)_2
```

results in the forced evaluation of bound pair 2:

```
(SQLIST (SUCC 0)_1)_2 => ... =>
((SQ (SUCC 0)_1)::(SQLIST (SUCC (SUCC 0)_1)_3)_4)_5
```

Thus:

```
TAIL (SQLIST (SUCC 0)_1)_2 => ... =>
TAIL ((SQ (SUCC 0)_1)::(SQLIST (SUCC (SUCC 0)_1)_3)_4)_5 =>... =>
(SQLIST (SUCC (SUCC 0)_1)_3)_4
```

leads to the selection of bound pair 4. Note that all occurrences of bound pair 2 were replaced by the new bound pair 5, so SQUARES is now associated with the list:

(SQ 0)::((SQ (SUCC 0)$_1$)::(SQLIST (SUCC (SUCC 0)$_1$)$_3$)$_4$)$_5$

The final level of selection:

HEAD (TAIL (TAIL SQUARES)) => ... =>

HEAD (SQLIST (SUCC (SUCC 0)$_1$)$_3$)$_4$

results in the forced evaluation of bound pair 4:

(SQLIST (SUCC (SUCC 0)$_1$)$_3$)$_4$ => ... =>

(SQ (SUCC (SUCC 0)$_1$)$_3$)$_6$::(SQLIST (SUCC (SUCC (SUCC 0)$_1$)$_3$)$_7$)$_8$

Occurrences of bound pair 4 are replaced by the new bound pair 8, so SQUARES is now:

(SQ 0)::(SQ (SUCC 0)$_1$)::(SQ (SUCC (SUCC 0)$_1$)$_3$)$_6$::
(SQLIST (SUCC (SUCC (SUCC 0)$_1$)$_3$)$_7$)$_8$

Thus, evaluation of:

HEAD (TAIL (TAIL SQUARES))

requires the selection of:

(SQ (SUCC (SUCC 0)$_1$)$_3$)$_6$

This requires the evaluation of:

(SUCC (SUCC 0)$_1$)$_3$

which in turn requires the evaluation of:

(SUCC 0)$_1$ => ... => 1

which replaces all occurrences of bound pair 1. Now, SQUARES is:

(SQ 0)::(SQ 1)::(SQ (SUCC 1)$_3$)$_6$::(SQLIST (SUCC (SUCC 1)$_3$)$_7$)$_8$

Thus evaluation of:

(SUCC (SUCC 0)$_1$)$_3$ => ... =>

(SUCC 1)$_3$

gives:

2

which replaces all occurrences of bound pair 3, so SQUARES is now:

(SQ 0)::(SQ 1)::(SQ 2)$_6$::(SQLIST (SUCC 2)$_7$)$_8$

Finally, evaluation of:

(SQ 2)$_6$

gives:

4

which replaces all occurrences of bound pair 6 so SQUARES is now:

(SQ 0)::(SQ 1)::4::(SQLIST (SUCC 2))

If we now try:

SQUARE 1 => ... =>

IFIND 1 SQUARES => ... =>

IFIND 0 (TAIL SQUARES) => ... =>

HEAD (TAIL SQUARES) ==

HEAD (TAIL ((SQ 0)::(SQ 1)$_9$::4::(SQLIST (SUCC 2)))) -> ... ->

HEAD ((SQ 1)$_9$::4::(SQLIST (SUCC 2))) => ... =>

(SQ 1)$_9$ => ... => 1

which replaces all occurrences of bound pair 9, so SQUARES is now:

(SQ 0)::1::4::(SQLIST (SUCC 2))

Thus, repeated list access evaluates more and more of the infinite list but avoids repetitive evaluation. Miranda is a fully lazy language. Thus, all data structures are evaluated lazily.

SUMMARY

- Normal order reduction may be less efficient than applicative order β reduction.
- Consistent applicative order β reduction with our conditional expression representation leads to non-termination.

- Applicative order evaluation can be delayed by various methods.
- The halting problem is unsolvable, but the Church-Rosser theorems suggest that normal order β reduction is most likely to terminate.
- Normal order β reduction enables the construction of infinite objects.
- Lazy evaluation is a way of combining the best aspects of normal and applicative order β reduction.

Lazy evaluation is summarized below.

Lazy evaluation

(1) Number every bound pair.

(2) To lazy evaluate (<function expression> <argument expression>)$_i$:

(a) Lazy evaluate <function expression> to <function value>.

(b) If <function value> is λ<name>.<body>
then replace all free occurrences of <name> in <body> with
<argument expression>
and renumber consistently all surrounding bound pairs
and replace all occurrences of
(<function expression> <argument expression>)$_i$
with the new <body>
and lazy evaluate the new <body>.

or:

(c) If <function value> is not a function
then lazy evaluate <argument expression> to
<argument value>
and replace all occurrences of
(<function expression> <argument expression>)$_i$
with (<function value> <argument value>)
and return (<function value> <argument value>).

EXERCISES

8.1 Evaluate the following expressions using normal order, applicative order and lazy evaluation. Explain any differences in the final result and the number of reductions in each case:

(a) λs.(s s) (λf.λa.(f a) λx.x λy.y)

(b) λx.λy.x λx.x (λs.(s s) λs.(s s))

(c) λa.(a a) (λf.λs.(f (s s)) λx.x)

Chapter 9
Functional programming in Standard ML

ML (Meta Language) is a general purpose language with a powerful functional subset. It is used mainly as a design and implementation tool for computing theory-based research and development. It is also used as a teaching language. ML is strongly typed with compile-time type checking. Function calls are evaluated in applicative order.

ML originated in the mid 1970s as a language for building proofs in Robin Milner's Logic for Computable Functions (LCF) computer-assisted formal reasoning system. Standard ML (SML) was developed in the early 1980s from ML with extensions from the Hope functional language. SML is one of the first programming languages to be based on well-defined theoretical foundations.

We will not give a full presentation of SML. Instead, we will concentrate on how SML relates to our approach to functional programming.

SML is defined in terms of a very simple **bare language** which is overlaid with standard **derived forms** to provide a higher level syntax. Here, we will just use these derived forms.

We will explain SML with examples. As the symbol -> is used in SML to represent the types of functions, we will follow SML system usage and show the result of evaluating an expression as:

```
- <expression>;
> <result>
```

9.1 Types

Types are central to SML. Every object and construct is typed. Unlike Pascal, types need not be made explicit but they must be capable of being deduced statically from a program. SML provides several standard types, for example booleans, integers, strings, lists and tuples which we will look at below. SML also has a variety of mechanisms for defining new types but we will not consider these here.

When representing objects, SML always displays types along with values:

```
<value> : <type>
```

For objects which do not have printable value representations, SML will still display the types. In particular, function values are displayed as:

```
fn : <type>
```

Types are described by type expressions. We will see how these are constructed as we discuss specific types.

9.1.1 Basic types – booleans, integers and strings

The type expression for a basic type is the type's **identifier**.

The boolean type has identifier:

```
bool
```

and values:

```
true  false
```

For example:

```
- true;
> true : bool
```

The integer type has identifier:

```
int
```

with positive and negative integer values, for example:

```
- 42;
> 42 : int
- ~84;
 ~84 : int
```

Note the use of ~ as the negative sign.

The string type has identifier:

```
string
```

String values are character sequences within "s, for example:

```
- "Is this a string?";
> "Is this a string?" : string
```

9.2 Lists

In SML, unlike LISP and our approach to λ calculus, a list must contain elements of the same type and end with the empty list. Thus, lists cannot be used to represent records with different type fields. Lists are written as ',' separated element sequences within '[' and ']'. For example:

```
[1,4,9,16,25]
["ant","beetle","caterpillar","dragonfly","earwig"]
```

There is an implied empty list at the end of a list. The empty list is:

```
[ ] or nil
```

The type expression for a list depends on the element type:

```
<element type> list
```

The first example above:

```
- [1,4,9,16,25];
> [1,4,9,16,25] : int list
```

is a list of integers. The second example above:

```
- ["ant","beetle","caterpillar","dragonfly","earwig"];
> ["ant","beetle","caterpillar","dragonfly","earwig"] : string list
```

is a list of strings.

Lists may be nested, for example:

```
- [[1,1],[2,8],[3,27],[4,64],[5,125]];
> [[1,1],[2,8],[3,27],[4,64],[5,125]] : (int list) list
```

is a list of integer lists. Note the use of ‘(’ and ‘)’ to structure the type expression.

9.3 Tuples

An ML **tuple**, like a Pascal RECORD, is a fixed length sequence of elements of different types, unlike a list which is a variable length sequence of elements of the same type. Tuples are written as ‘,’ separated sequences within ‘(’ and ‘)’. For example, we might represent a stock control record from Chapter 7 as:

```
("VDUs",250,120)
```

Tuples may be nested. For example, we might represent a telephone directory entry from Chapter 7 as:

```
(("Anna","Able"),"Accounts",101)
```

A tuple's type is represented by its elements' types separated by ‘*’:

```
<element1 type> * <element2 type> * ...
```

For example:

```
- ("VDUs",250,120);
> ("VDUs",250,120) : string * int * int
```

is a tuple consisting of two integers and a string.

```
- (("Anna","Able"),"Accounts",101);
> (("Anna","Able"),"Accounts",101) : (string * string) * string * int
```

is a tuple consisting of a tuple consisting of two strings, a string and an integer. These may be used to build tuple lists, for example, a stock control list:

```
- [("VDUs",250,120),
   ("mice",155,170),
   ("printers",43,20)];
> [("VDUs",250,120),
  ("mice",155,170),
  ("printers",43,20)] : (string * int * int) list
```

or a telephone directory:

```
- [(("Anna","Able"),"Accounts",101),
   (("Betty","Baker"),"Boiler room",102),
   (("Cloe","Charlie"),"Customer orders",103)];
> [(("Anna","Able"],"Accounts",101),
  (("Betty","Baker"),"Boiler room",102),
  (("Cloe","Charlie"),"Customer orders",103)] :
   ((string * string) * string * int) list
```

Note that if a tuple is defined with expression elements, then those expressions are evaluated from left to right. Thus, as tuples are used to specify bound variables for uncurried functions, such functions have a defined actual parameter evaluation order.

9.4 Function types and expressions

A function uses values in an argument **domain** to produce a final value in a result **range**. In SML, a function's type is characterized by its domain and range types:

```
fn : <domain type> -> <range type>
```

Note the use of '->' to indicate a function's domain/range mapping. Tuples are normally used to enable uncurried functions with multiple bound variables.

In SML, as in λ calculus and LISP, expressions are usually based on prefix notation function applications with the function preceding the arguments:

```
<function expression> <argument expression>
```

Function applications are evaluated in applicative order. Note that function applications need not be explicitly bracketed, but brackets should be used around arguments to avoid ambiguity. SML enables uncurried binary functions to be used as infix operators so the function name may appear in between the two arguments. They are then typed as if they had tuples for arguments. We will not consider this further here.

Similarly, many standard binary functions are provided as infix operators. They may be treated as prefix functions on tuples by preceding them with:

```
op
```

We will look at this in more detail in the following sections.

9.5 Standard functions

Let us consider the following standard functions.

9.5.1 Boolean standard functions

The boolean negation function:

```
not
```

returns the negation of its boolean argument, for example:

```
- not true;
> false : bool
```

Thus, not's type is:

```
- not;
> fn : bool -> bool
```

Conjunction and disjunction are provided through the sequential infix operators:

```
andalso orelse
```

in the derived syntax, for example:

```
– true orelse false;
> true : bool

– true andalso false;
> false : bool
```

SML systems may not be able to display these operators' types, but they are effectively:

```
fn : (bool * bool) –> bool
```

as they both take two boolean arguments, which are treated as a:

```
bool * bool
```

tuple for infix syntax, and return a boolean result.

9.5.2 Numeric standard functions and operator overloading

SML provides real numbers as well as integers. However, as in many other languages, the same operators are used for both even though they are distinct types. This use of the same operator with different types is known as **operator overloading.**

The addition, subtraction and multiplication infix operators are:

```
+ – *
```

SML systems may not display their types because they are overloaded. SML literature uses the invented type:

```
num
```

to indicate both integer and real, so these operators types might be:

```
fn : (num * num) –> num
```

as they take two numeric arguments, with infix syntax for a tuple, and return a numeric result. Note that for each operator both arguments must be the same type.

The infix operator:

```
div
```

is for integer division. We can use op to convert it to prefix form to display its type:

```
- op div;
> fn : (int * int) -> int
```

Arithmetic expressions are built from these operators with the brackets '(' and ')', for example:

```
- 6 * 7 div (7 - 4) + 28;
> 42 : int
```

Note that there is no strict bracketing. The usual precedence:

```
( )
```

before:

```
div *
```

before:

```
+ -
```

applies.

The numeric negation operator is:

```
~
```

again with effective type:

```
fn : num -> num
```

as it is overloaded for use with integers and reals.

9.5.3 String standard functions

The binary infix operator:

```
^
```

concatenates two strings together:

```
- op^;
> fn : (string * string) -> string
```

For example:

```
- "Happy"^" birthday!";
> "Happy birthday!" : string
```

The operator:

```
size
```

returns the size of a string:

```
- size;
> fn : string -> int
```

For example:

```
- size "hello";
> 5 : int
```

Standard functions for turning strings into string lists are discussed below.

9.5.4 List standard functions

In SML, list operations apply to lists of any types. In SML, an unknown type is denoted by a single letter name preceded by a prime "'", for example:

```
'a 'b 'c
```

Thus, we can refer to a list of arbitrary typed objects as having type:

```
'a list
```

In SML, lists are accessed by the head and tail operators:

```
hd tl
```

The head operator returns the head object with type:

```
'a
```

from an arbitrary typed list. Thus, hd is of type:

```
- hd;
> fn : ('a list) -> 'a
```

For example:

```
- hd [1,2,3,4,5];
> 1 : int
```

Similarly, the tail operator returns the tail with type:

```
'a list
```

from an arbitrary type list. Thus, tl is of type:

```
- tl;
> fn : ('a list) -> ('a list)
```

For example:

```
- tl ["alpha","beta","gamma","delta","epsilon"];
> ["beta","gamma","delta","epsilon"] : string list
```

The infix list concatenation operator is:

```
::
```

Given an object and a list of the same type of object, '::' returns a new list with the object in the head and the object list in the tail. Thus, '::' has type:

```
- op ::;
> ('a * ('a list)) -> ('a list)
```

For example:

```
- 0::[1,2,3,4,5];
> [0,1,2,3,4,5] : int list
```

The operators hd, tl and '::' are said to be **polymorphic** because they apply to a list of any type of object. We will look at polymorphism in slightly more detail later.

9.5.5 Characters, strings and lists

SML does not provide a separate character type. Instead, a character is a one letter string. The standard function:

```
ord
```

converts a single character string to the equivalent ASCII code value:

```
- ord;
> fn : string -> int
```

For example:

```
- ord "a";
> 97 : int
```

Similarly, the standard function:

```
chr
```

converts an integer ASCII value into the equivalent single character string:

```
- chr;
> fn : int -> string
```

For example:

```
- chr 54;
> "6" : string
```

In order to access the individual characters making up a string, it must be unpacked into a list of single character strings. The standard function:

```
explode
```

does this:

```
- explode;
> fn : string -> (string list)
```

For example:

```
- explode "hello";
> ["h","e","l","l","o"] : string list
```

Similarly, the standard function:

```
implode
```

converts a list of strings to a single string:

```
- implode;
> fn : (string list) -> string
```

For example:

```
- implode ["Time ","for ","tea?"];
> "Time for tea?" : string
```

Note that implode will join strings of any length.

9.6 Comparison operators

SML provides a variety of overloaded infix comparison operators. Equality and inequality are tested with:

```
= <>
```

and may be used with booleans, integers, strings, lists and tuples. The less than, less than or equal, greater than or equal and greater than operators:

```
< <= >= >
```

may be used with numbers and strings. For strings, they test for alphabetical order, for example:

```
- "honey" < "orange";
> true : bool
```

SML systems may not display these operators' types because they are overloaded.

For all these operators, both arguments must be of the same type.

9.7 Functions

Functions have the form:

```
fn <bound variables> => <expression>
```

A bound variable is known as an **alphanumeric identifier** and consists of one or more letters, digits and '_'s starting with a letter, for example:

```
oxymoron Home_on_the_range Highway61
```

A function's bound variable may be a single bound variable or a tuple of bound variable elements. For example:

```
- fn x => x+1;
> fn : int -> int
```

increments its argument. Note that SML deduces that the domain and range are int because '+' is used with the int argument 1.

As a further example:

```
- fn x => fn y => not (x orelse y);
> fn : bool -> (bool -> bool)
```

is the boolean implication function. Note that orelse has a boolean tuple domain so x and y must both be boolean. Similarly, not returns a boolean, so the inner function:

```
fn y => not (x orelse y)
```

has type:

```
bool -> bool
```

Hence, the whole function has type:

```
fn : bool -> (bool -> bool)
```

This might have been written with a tuple domain:

```
- fn (x,y) => not (x orelse y);
> fn : (bool * bool) -> bool
```

9.8 Making bound variables' types explicit

Suppose we try to define a squaring function:

```
fn x => x*x
```

Because '*' is overloaded, SML cannot deduce x's type and will reject this function.

Domain types may be made explicit by following each bound variable with its type. Thus for a single bound variable:

```
(<bound variable> : <type>)
```

is used. For example, an integer squaring function may be defined by:

```
- fn (x:int) => x*x;
> fn : int -> int
```

For a tuple of bound variables:

```
(<bound variable1> : <type1>, <bound variable2> : <type2>, ... )
```

is used. For example, we might define the sum of squares function as:

```
- fn (x:int,y:int) => x*x+y*y;
> fn : (int * int) -> int
```

It is thought to be 'good practice' to make all bound variables' types explicit. This is supposed to make it easier to read functions and to ensure that types are consistent. However, without care, type expressions can become unmanageably long. SML provides ways to name complex types which we will consider in a later section.

9.9 Definitions

Global definitions may be established with:

```
val <name> = <expression>
```

For example:

```
- val sq = fn (x:int) => x*x;
> val sq = fn : int -> int
- val sum_sq = fn (x:int,y:int) => x*x+y*y;
> val sum_sq = fn : (int * int) -> int
```

Note that the SML system acknowledges the definition by displaying the defined name and the expression's value and/or type.

Defined names may be used in subsequent expressions, for example:

```
- sq 3;
> 9 : int
```

and in subsequent definitions, for example:

```
- val sum_sq = fn (x:int,y:int) => (sq x)+(sq y);
> val sum_sq = fn : (int * int) -> int
```

9.10 Conditional expressions

The SML conditional expression has the form:

```
if <expression1>
then <expression2>
else <expression3>
```

The first expression must return a boolean, and the option expressions <expression2> and <expression3> must have the same type. Note that SML conditional expressions are evaluated in normal order. For example, to find the larger of two integers:

```
- val max = fn (x:int,y:int) => if x>y
                                  then x
                                  else y;
> val max = fn : (int * int) -> int
```

Or, to define sequential boolean implication:

```
- val imp = fn (x,y) => if x
                         then y
                         else true;
> val imp = fn : (bool * bool) -> bool
```

9.11 Recursion and function definitions

To define recursive functions, the defined name is preceded by:

```
rec
```

For example, to find the length of an integer list:

```
- val rec length = fn (l:int list) => if l = []
                                       then 0
                                       else 1+(length (tl l));
> val length = fn : (int list) -> int
```

As with our λ calculus notation, there is a shortened form for function definitions. Instead of val:

```
fun
```

is used to introduce the definition, the fn is dropped, the bound variables are moved to the left of the '=' and the '=>' is dropped. For recursive definitions, the rec is dropped. Thus:

```
fun <name> <bound variables> = <expression> ==
val rec <name> = fn <bound variables> => <expression>
```

For example, to square all the values in an integer list:

```
- fun squarel (l:int list) =
    if l=[ ]
    then [ ]
    else ((hd l)*(hd l))::(squarel (tl l));
> fun squarel = fn : (int list) -> (int list)
```

Or to insert a string into an ordered string list:

```
- fun sinsert (s:string,l:string list) =
    if l = [ ]
    then [s]
    else
     if s < (hd l)
     then s::l
     else (hd l)::(sinsert (s,(tl l)));
> val sinsert = fn : (string * (string list)) -> (string list)
```

9.12 Tuple selection

Tuple elements are selected by defining functions with appropriate bound variable tuples. For example, to select the name, department and phone number from a telephone directory entry tuple:

```
- fun tname (n:(string * string),d:string,p:int) = n;
> val tname = fn : ((string * string) * string * int) ->
                   (string * string)

- fun tdept (n:(string * string),d:string,p:int) = d;
> val tdept = fn : ((string * string) * string * int) -> string

- fun tno (n:(string * string),d:string,p:int) = p;
> val tno = fn : ((string * string) * string * int) -> int
```

To avoid writing out bound variables which are not used in the function body, SML provides the **wild card** variable:

```
-
```

which behaves like a nameless variable of arbitrary type. For example, we could rewrite the above examples as:

```
- fun tname (n:(string * string),_,_) = n;
> val tname = fn : ((string * string) * 'a * 'b) -> (string * string)

- tname (("Anna","Able"),"Accounts",123);
> ("Anna","Able") : (string * string)

- fun tdept (_,d:string,_) = d;
> val tdept = fn : ('a * string *'b) -> string

- tdept (("Anna","Able"),"Accounts",123);
> "Accounts" : string

- fun tno (_,_,p:int) = p;
> val tno = fn : ('a * 'b * int) -> int

- tno (("Anna","Able"),"Accounts",123);
> 123 : int
```

Note that SML uses 'a and 'b to stand for possibly distinct unknown types.

For nested tuple selection, nested bound variable tuples are used. For example, to select the forename and surname from a telephone directory entry:

```
- fun fname ((f:string,_),_,_) = f;
> val fname = fn : ((string * 'a) * 'b * 'c) -> string

- fname (("Anna","Able"),"Accounts",123);
> "Anna" : string

- fun sname ((_,s:string),_,_) = s;
> val fname = fn : (('a * string) * 'b * 'c) -> string

- sname (("Anna","Able"),"Accounts",123);
> "Able" : string
```

9.13 Pattern matching

SML functions may be defined with bound variable patterns using constants and constructors as well as variables. For example, the head and tail selector functions for integer lists might be defined by:

```
- fun ihd ((h:int)::(t:int list)) = h;
> val ihd = fn : (int list) -> int

- fun itl ((h:int)::(t:int list)) = t;
> val itl = fn : (int list) -> (int list)
```

Note the use of the bound variable pattern:

```
((h:int)::(t:int list))
```

with the list constructor '::'. Note also that this function will crash with an empty list argument as the pattern match will fail.

It is common SML practice to use case-style function definitions with pattern matching rather than conditional expressions in a function's body. These are known as **clausal form** definitions. The general form is:

```
fun <name> <pattern1> = <expression1> |
    <name> <pattern2> = <expression2> |
    ...
    <name> <patternN> = <expressionN>
```

Here, each:

```
<name> <patternI> = <expressionI>
```

defines a case. Note that the order of the cases is significant.

When a case defined function is applied to an argument, each pattern is matched against the argument in turn, from first to last, until one succeeds. The value of the corresponding expression is then returned. Consider the following examples: we might construct a function to return the capital of a Scandinavian country as a sequence of constant cases:

```
- fun capital "Denmark" = "Copenhagen" |
      capital "Finland" = "Helsinki" |
      capital "Norway" = "Oslo" |
      capital "Sweden" = "Stockholm" |
      capital _ = "not in Scandinavia";
> val capital = fn : string -> string
```

Here, an argument is compared with constants until the last case, where it is matched with the wild card variable.

We might also redefine the integer list length function in terms of a base case for the empty list and a recursive case for a non-empty list:

```
- fun length [] = 0 |
      length (_::(t:int list)) = 1+(length t);
> val length = fn : (int list) -> int
```

Here an argument is compared with the empty list in the first case or split into its head and tail in the second. The head is matched with the wild card variable and lost.

We might also generate a list of the first n cubes, with a base case for when n is 0, and a recursive case for positive n:

```
- fun cubes 0 = [0] |
      cubes (n:int) = (n*n*n)::(cubes (n-1));
> val cubes = fn : int -> (int list)
```

Here, an argument is compared with 0 in the first case or associated with the bound variable n in the second.

We might find the ith element in a string list with a base case which fails for an empty list, a base case which returns the head of the list when i is 0 and a recursive case for positive i with a non-empty list:

```
- fun sifind _ [] = "can't find it" |
      sifind 0 ((h:string)::_) = h |
      sifind (i:int) (_::(t:string list)) = sifind (i-1) t;
> val sfiind = fn : int -> ((string list) -> string)
```

Here, the integer argument is matched with the wild card variable in the first case, compared with 0 in the second and associated with the bound variable i in the third. Similarly, the list argument is compared with the empty list in the first case and split into its head and tail in the second and third. In the second case, the tail is matched with with the wild card variable and lost. In the third case, the head is matched with the wild card variable and lost. Note that this is a curried function.

Patterns may also be used to specify nameless functions.

9.14 Local definitions

SML uses the let ... in ... notation for local definitions:

```
let val <name> = <expression1>
in <expression2>
end
```

This evaluates <expression2> with <name> associated with <expression1>.

For function definitions:

```
let fun <name> <pattern> = <expression1>
in <expression2>
end
```

and the corresponding case form is used. For example, to find the number of unique i element combinations from n elements:

```
n!/(i!*(n-i)!)
```

using a local factorial function:

```
- fun comb (i:int) (n:int) =
      let fun fac 0 = 1 |
              fac n = n*(fac (n-1))
      in (fac n) div ((fac i)*(fac (n - i)))
      end;
> val comb = fn : int -> (int -> int)
```

9.15 Type expressions and abbreviated types

We will now be a bit more formal about types in SML. We specify a variable's type with a **type expression**. Type expressions are built from **type constructors** like int, bool, string and list. So far, a type expression may be a single type constructor, a type variable, a function type, a product type, a bracketed type expression or a type variable preceding a type constructor. SML also enables the use of **abbreviated types** to name type expressions. A name may be associated with a type expression using a **type binding** of the form:

```
type <abbreviation> = <type expression>
```

The <abbreviation> is an identifier which may be used in subsequent type expressions. For example, in the telephone directory example, we might use abbreviations to simplify the types used in a directory entry:

```
- type forename = string;
> type forename = string

- type surname = string;
> type surname = string

- type person = forename * surname;
> type person = forename * surname

- type department = string;
> type department = string

- type extension = int;
> type extension = int

- type entry = person * department * extension;
> type entry = person * department * extension
```

New type constructors may be used in subsequent expressions in the same way as predefined types. Note that a new type constructor is syntactically equivalent to its defining expression. Thus, if we define:

```
- type whole_numb = int;
> type whole_numb = int

- type integer = int;
> type integer = int
```

then values of type whole_numb, integer and int may be used in the same places without causing type errors. This form of type binding disguises a longer type expression. Type abbreviations may also be parameterized but we will not consider that further here.

9.16 Type variables and polymorphism

SML, like Pascal, is strongly typed. All types must be determinable by static analysis of a program. Pascal is particularly restrictive because there is no means of referring to a type in general. SML, however, allows generalization through the use of type variables in type expressions where particular types are not significant.

A type variable starts with a '' and usually has only one letter, for example:

```
'a 'b 'c
```

We have already seen the use of type variables to describe the standard list functions' types and the use of the wild card variable. With strong typing but without type variables, generalized functions cannot be described. In Pascal, for example, it is not possible to write general purpose procedures or functions to process arrays of arbitrary types. In SML, though, the list type is generalized through a type variable to be element type independent.

Above, we described a variety of functions with specific types. Let us now look at how we can use type variables to provide more general definitions. Consider the following examples: the head and tail list selector functions might be defined as:

```
- fun hd (h::t) = h;
> val hd = fn : ('a list) -> 'a

- fun tl (h::t) = t;
> val tl = fn : ('a list) -> ('a list)
```

Here, in the pattern:

```
(h::t)
```

there are no explicit types. SML 'knows' that '::' is a constructor for lists of any type element provided the head and tail element type have the same type. Thus, if '::' is:

```
('a * ('a list)) -> ('a list)
```

then h must be 'a and t must be 'a list. We do not need to specify types here because list construction and selection is type independent. Note that we could have used a wild card variable for t in hd and for h in tl.

We can also define general functions to select the elements of a three place tuple:

```
- fun first (x,y,z) = x;
> val first = fn : ('a * 'b * 'c) -> 'a

- fun second (x,y,z) = y;
> val second = fn : ('a * 'b * 'c) -> 'b

- fun third (x,y,z) = z;
> val third = fn : ('a * 'b * 'c) -> 'c
```

Here, in the pattern:

```
(x,y,z)
```

there are no explicit types. Hence, SML assigns the types 'a to x, 'b to y and 'c to z. Here, the element types are not significant. For selection, all that matters is their relative positions within the tuple. Note that we could have used wild cards for y and z in first, for x and z in second, and for x and y in third.

We can also define a general purpose list length function:

```
- fun length [] = 0 |
      length (h::t) = 1+(length t);
> val length = fn : ('a list) -> int
```

There are no explicit types in the pattern:

```
(h::t)
```

and this pattern is consistent if h is 'a, and t is 'a list, as (h::t) is then 'a list. This is also consistent with the use of t as the argument in the recursive call to length. Here again, the element types are irrelevant for purely structural manipulations.

This approach can also be used to define type independent functions which are later made type specific. For example, we might try to define a general purpose list insertion function as:

```
- fun insert i [ ] = [i] |
      insert i (h::t) =
       if i<h
       then i::h::t
       else h::(insert i t);
```

but this is incorrect although the bound variable typing is consistent if i and h are both 'a and t is 'a list. The problem lies with the use of the comparison operator '<'. This is overloaded so its arguments' types must be made explicit. We could get round this by abstracting for the comparison:

```
- fun insert _ i [ ] = [i] |
      insert comp i (h::t) =
       if comp (i,h)
       then i::h::t
       else h::(insert comp i t);
> val insert = (('a * 'a) -> bool) ->
               ('a -> (('a list) -> ('a list)))
```

Here, comp needs an ('a * 'a) argument to be consistent in insert and must return a bool to satisfy its use in the if.

Different typed comparison functions may be used to construct different typed insertion functions. For example, we could construct a string insertion function through partial application by passing a string comparison function:

```
fn (s1:string,s2:string) => s1<s2
```

to replace comp in insert:

```
- val sinsert = insert (fn (s1:string,s2:string) => s1<s2);
> val sinsert = fn : string -> ((string list) -> (string list))
```

Here, the comparison function is:

```
(string * string) -> bool
```

so 'a must be string in the rest of the function's type.

As a further example, we could construct an integer insertion function through partial application by passing an integer comparison function:

```
fn (i1:integer,i2:integer) => i1<i2
```

to replace comp in insert:

```
- val iinsert = insert (fn (i1:int,i2:int) => i1<i2);
> val iinsert = fn : int -> ((int list) -> (int list))
```

Now, the comparison function is:

```
(int * int) -> bool
```

so 'a must be int in the rest of the function.

Functions which are defined for generalized types are said to be **polymorphic** because they have many forms. Polymorphic typing gives substantial power to a programming language and a great deal of research and development has gone into its theory and practice. There are several forms of polymorphism. Strachey distinguishes *ad hoc* polymorphism through operator overloading from 'parameterized' polymorphism, through abstraction over types. Cardelli distinguishes 'explicit' parameterized polymorphism, where the types are themselves objects, from the weaker 'implicit' polymorphism where type variables are used in type expressions but types are not themselves objects, as in SML. Milner first made type polymorphism in functional languages practical with his early ML for LCF. This introduced polymorphic type checking where types are deduced from type expressions and variable use. Hope and Miranda also have implicit parameterized polymorphic type checking.

9.17 New types

A new **concrete** type may be introduced by a **datatype binding**. This is used to define a new type's constituent values recursively by:

(1) Listing base values explicitly.
(2) Defining structured values in terms of base values and other structured values.

The binding introduces new **type constructors** which are used to build new values of that datatype. They are also used to identify and manipulate such values.

At its most simple, a datatype binding takes the form:

```
datatype <constructor> = <constructor1> |
                         <constructor2> |
                         ...
                         <constructorN>
```

which defines the base values of type <constructor>, an identifier, to be the type constructor identifiers <constructor1> or <constructor2> etc. For example, we could define the type bool with:

```
- datatype bool = true | false;
> datatype bool = true | false
  con true = true : bool
  con false = false : bool
```

This defines the constructors true and false for the new type bool. In effect, this specifies that an object of type bool may have either the value true or the value false. An equality test for bool is also defined so that the values true and false may be tested explicitly.

Consider a traffic light in Great Britain, which goes through the stages red, red and amber, green, amber and back to red:

```
- datatype traffic_light = red | red_amber | green | amber;
> datatype traffic_light = red | red_amber | green | amber
  con red = red : traffic_light
  con red_amber = red_amber : traffic_light
  con green = green : traffic_light
  con amber = amber : traffic_light
```

This defines the data type traffic_light with the constructors red, red_amber, green and amber. In effect, red, red_amber, green and amber are the values of the new type: traffic_light. An equality test for traffic_light values is also defined. For example, we can now define a function to change a traffic light from one stage to the next:

```
- fun change red       = red_amber |
      change red_amber = green     |
      change green     = amber     |
      change amber     = red;
> val change = fn : traffic_light -> traffic_light
```

The datatype binding is also used to define structured concrete types. The binding form is extended to:

```
datatype <constructor> = <constructor1> of <type expression1> |
                         <constructor2> of <type expression2> |
                         ...
                         <constructorN> of <type expressionN>
```

where the extension of <type expression> is optional. This specifies a new type:

```
<constructor>
```

with values of the form:

```
<constructor1>(<value for <type expression1>>)
<constructor2>(<value for <type expression2>>)
etc.
```

<constructor1>, <constructor2>, etc. are functions which build structured values of type <constructor>. For example, integer lists might be defined by:

```
- datatype intlist = intnil | intcons of int * intlist;
> datatype intlist = intnil | intcons of int * intlist
  con intnil = intnil : intlist
  con intcons = fn : (int * intlist) -> intlist
```

Now, intnil is an intlist and values of the form intcons(<int value>, <intlist value>) are intlist. That is, intcons is a function which builds an intlist from an int and an intlist. For example:

```
- intcons(1,intnil);
> intcons(1,intnil) : intlist
- intcons(1,intcons(2,intnil));
> intcons(1,intcons(2,intnil)) : intlist
- intcons(1,intcons(2,intcons(3,intnil)));
> intcons(1,intcons(2,intcons(3,intnil))) : intlist
```

A datatype constructor may be preceded by a type variable to parameterize the datatype. For example, SML lists might be defined by:

```
- datatype 'a list = lnil | cons of 'a * ('a list);
> datatype 'a list = lnil | cons of 'a * ('a list)
  con lnil : 'a list
  con cons = fn : ('a * 'a list) -> ('a list)
```

This defines cons as a prefix constructor.

Type variables in datatype definitions may be set to other types in subsequent type expressions. For example, in:

```
- type intlist = int list;
> type intlist = int list
```

the type variable 'a is set to int to use intlist to name an integer list type.

SML systems will also deduce the intended type when the constructor from a parameterized data type is used with consistent values. Thus, the following are all string lists:

```
- cons("ant",lnil);
> cons("ant",lnil) : string list

- cons("ant",cons("bee",lnil));
> cons("ant",cons("bee",lnil)) : string list

- cons("ant",cons("bee",cons("caterpillar",lnil)));
> cons("ant",cons("bee",cons("caterpillar",lnil))) : string list
```

Structured datatype values may also be used in patterns with typed variables and constructors in the tuple following the datatype constructor. It is usual to have separate patterned definitions for base and for structured values. For example, to sum the elements of an intlist:

```
- fun sum intnil = 0 |
      sum (intcons(x:int,y:intlist)) = x + (sum y);
> val sum = fn : intlist -> int

- sum (intcons(9,intcons(8,intcons(7,intnil))));
> 24 : int
```

Or, to join the elements of a string list:

```
- fun join lnil = "" |
      join (cons(s:string,l:(string list))) = s^join l;
> val join = fn : (string list) -> string

- join (cons("here",cons("we",cons("go",lnil))));
> "herewego" : string
```

Note that existing types cannot be used as base types directly. For example, we might try to define a general number type as:

```
- datatype number = int | real;
> datatype number = int | real
  con int = int : number
  con real = int : number
```

but this defines the new type number with base constructors int and real as if they were simple identifiers for constant values instead of types. A structured constructor must be used to incorporate existing types into a new type, that is:

```
- datatype number = intnumb of int | realnumb of real;
> datatype number = intnumb of int | realnumb of real
  con intnumb = fn : int -> number
  con realnumb = fn : real -> number
```

For example:

```
- intnumb(3);
> intnumb(3) : number

- realnumb(3.3);
> realnumb(3.3) : number
```

Note that structure matching must now be used to extract the values from this structured type:

```
- fun ivalue (intnumb(n:int)) = n;
> val ivalue = fn : number -> int

- fun rvalue (realnumb(r:real)) = r;
> val rvalue = fn : number -> real
```

So, for example:

```
- ivalue (intnumb(3));
> 3 : int

- rvalue (realnumb(3.3));
> 3.3 : real
```

9.18 Trees

We looked at tree construction in Chapter 7. We will now see how SML concrete datatypes may be used to construct trees. First of all we will consider binary integer trees.

To recap: a binary integer tree is either empty or it is a node consisting of an integer value, a left subtree and a right subtree. Thus, we can define a corresponding datatype:

```
- datatype inttree = empty | node of int * inttree * inttree;
> datatype inttree = empty | node of int * inttree * inttree
  con empty = empty : inttree
  con node = fn : (int * inttree * inttree) -> inttree
```

To add an integer to an integer binary tree, if the tree is empty, then form a new node with the integer as value and empty left and right subtrees.

Otherwise, if the integer comes before the root node value, then add it to the left subtree and if it comes after the root node value, then add it to the right subtree:

```
- fun tadd (v:int) empty = node(v,empty,empty) |
      tadd (v:int) (node(nv:int,l:inttree,r:inttree)) =
        if v < nv
        then node(nv,tadd v l,r)
        else node(nv,l,tadd v r);
> val tadd = fn : int -> (inttree -> inttree)
```

For example:

```
- val root = empty;
> val root = empty : inttree

- val root = tadd 5 root;
> val root = node(5,empty,empty) : inttree

- val root = tadd 3 root;
> val root = node(5,
                  node(3,empty,empty),
                  empty) : inttree

- val root = tadd 7 root;
> val root = node(5,
                  node(3,empty,empty),
                  node(7,empty,empty)) : inttree

- val root = tadd 2 root;
> val root = node(5,
                  node(3,
                       node(2,empty,empty),
                       empty),
                  node(7,empty,empty)) : inttree

- val root = tadd 4 root;
> val root = node(5,
                  node(3,
                       node(2,empty,empty),
                       node(4,empty,empty)),
                  node(7,empty,empty)) : inttree

- val root = tadd 9 root;
> val root = node(5,
                  node(3,
                       node(2,empty,empty),
                       node(4,empty,empty)),
```

```
                    node(7,
                         empty,
                         node(9,empty,empty))) : inttree
```

Given an integer binary tree, to construct an ordered list of node values: if the tree is empty, then return the empty list; otherwise, traverse the left subtree, pick up the root node value and traverse the right subtree:

```
- fun traverse empty = [] |
      traverse (node(v:int,l:inttree,r:inttree)) =
       append (traverse l) (v::traverse r);
> val traverse = fn : inttree -> (int list)
```

For example:

```
- traverse root;
> [2,3,4,5,7,9] : int list
```

We can rewrite the above datatype to specify trees of polymorphic type by abstracting with the type variable 'a:

```
- datatype 'a tree = empty | node of 'a * ('a tree) * ('a tree);
> datatype 'a tree = empty | node of 'a * ('a tree) * ('a tree)
  con empty = empty : ('a tree)
  con node = fn : ('a * ('a tree) ('a tree)) -> ('a tree)
```

Similarly, we can define polymorphic versions of add:

```
- fun tadd _ (v:'a) empty = node(v,empty,empty) |
      tadd (less:'a -> ('a -> bool))
           (v:'a)
           (node(nv:'a,l:'a tree,r:'a tree)) =
       if less v nv
       then node(nv,tadd less v l,r)
       else node(nv,l,tadd less v r);
> val tadd = fn : ('a -> ('a -> bool)) ->
                  ('a -> (('a tree) -> ('a tree)))
```

and traverse:

```
- fun traverse empty = [] |
      traverse (node(v:'a,l:'a tree,r:'a tree)) =
       append (traverse l) (v::traverse r);
> val traverse = fn : ('a tree) -> ('a list)
```

Note the use of the bound variable less in add to generalize the comparison between the new value and the node value.

9.19 λ calculus in SML

We can use SML to represent many pure λ functions directly. For example, the identity function is:

```
- fn x => x;
> fn : 'a -> 'a
```

Note that this is a polymorphic function from the domain 'a to the same range 'a. Let us apply the identity function to itself:

```
- (fn x => x) (fn x => x);
> fn : 'a -> 'a
```

Alas, SML will not display nameless functions. For example, the function application function is:

```
- fn f => fn x => (f x);
> fn : ('a -> 'b) -> ('a -> 'b)
```

This is another polymorphic function. Here, f is used as a function but its type is not specified, so it might be 'a -> 'b for arbitrary domain 'a and arbitrary range 'b. f is applied to x, so x must be 'a. The whole function returns the result of applying f to x which is of type 'b. Let us use this function to apply the identity function to itself:

```
- (fn f => fn x => (f x)) (fn x => x) (fn x => x);
> fn : 'a -> 'a
```

Once again, SML will not display the resulting function. Using global definitions does not help here:

```
- val identity = fn x => x;
> val identity = fn : 'a -> 'a

- identity identity;
> fn : 'a -> 'a

- val apply = fn f => fn x => (f x);
> val apply = fn : ('a -> 'b) -> ('a -> 'b)

- apply identity identity;
> fn : 'a -> 'a
```

as applicative order evaluation replaces name arguments with values.

Some of our λ functions cannot be represented directly in SML as the type system will not allow self-application. For example, in:

```
fn s => (s s)
```

there is a type inconsistency in the function body:

```
(s s)
```

Here, the s in the function position is untyped so it might be 'a -> 'b. Thus, the s in the argument position should be 'a, but this clashes with the type for the s in the function position!

9.20 Other features

There are many aspects of SML which we cannot cover here. Most important are abstract type construction and modularization techniques and the use of exceptions to change control flow, in particular to trap errors. SML also provides imperative constructs for assignment, I/O and iteration.

SUMMARY

- Standard ML (SML) is a general purpose language which is well suited to functional programming.
- Algorithms from preceding chapters can be implemented in SML.
- Some pure λ functions cannot be represented in SML.

EXERCISES

9.1 Write and test functions to:

(a) Find y^3 given integer y.

(b) Find x implies y from x implies y == not x or y given x and y. The function implies should be prefix.

(c) Find the smallest of the integers a, b and c.

(d) Join strings s1 and s2 together in descending alphabetical order.

(e) Find the shorter of strings s1 and s2.

9.2 Write and test functions to:

(a) Find the sum of the integers between 1 and n.

(b) Find the sum of the integers between m and n.

(c) Repeat a string s integer n times.

9.3 Write and test functions to:

(a) Count the number of negative integers in a list l.

(b) Count how often a given string s occurs in a list l.

(c) Construct a list of all the integers in a list l which are greater than a given value v.

(d) Merge two sorted string lists s1 and s2. For example:

```
- smerge ["a","d","e"] ["b","c","f","g"];
> ["a","b","c","d","e","f","g"] : string list
```

(e) Use smerge from (iv) above to construct a single sorted string list from a list of sorted string lists. For example:

```
- slmerge [["a","c","i","j"],["b","d","f"],["e","g","h","k"]];
> ["a","b","c","d","e","f","g","h","i","j","k"] : string list
```

(f) Process a list of stock records represented as tuples of:

```
item name     – string
no. in stock  – integer
reorder level – integer
```

to:

(i) Construct a list of those stock records with the number in stock less than the reorder level. For example:

```
- check [("RAM",9,10),("ROM",12,10),("PROM",20,21)];
> [("RAM",9,10),("PROM",20,21)] : (string * int * int) list
```

(ii) Update a list of stock records from a list of update records represented as tuples of:

```
item name – string
update no. – integer
```

by constructing a new stock list of records with the number in stock increased by the update number. The update records may be in any order. There may not be update records for all stock items. There may be more than one update record for any stock item. For example:

```
- update [("RAM",9,10),("ROM",12,10),("PROM",20,21)]
         [("PROM",15),("RAM",12),("PROM",15)];
> [("RAM",21,10),("ROM",12,10),("PROM",50,21)] :
  (string * int * int) list
```

9.4 Write functions to:

(a) Extract the leftmost n letters from string s:

```
- left 4 "goodbye";
> "good" : string
```

(b) Extract the rightmost n letters from string s:

```
- right 3 "goodbye";
> "bye" : string
```

(c) Extract n letters starting with the lth letter from string s:

```
- middle 2 5 "goodbye";
> "by" : string
```

(d) Find the position of the first occurrence of string s1 in string s2:

```
- find "by" "goodbye";
> 5 : int
```

9.5 The train travelling east from Glasgow Queen Street to Edinburgh Waverly passes through Bishopbriggs, Lenzie, Croy, Polmont, Falkirk High, Linlithgow and Edinburgh Haymarket. These stations might be represented by the data type:

```
datatype station = Queen_Street | Bishopbriggs |
                   Lenzie | Croy |
                   Polmont | Falkirk_High |
                   Linlithgow | Haymarket | Waverly;
```

Write functions which return the station to the east or the west of a given station, for example:

```
- east Croy;
> Polmont : station

- west Croy;
> Lenzie : station
```

9.6 The data type:

```
datatype exp = add of exp * exp |
               diff of exp * exp |
               mult of exp * exp |
               quot of exp * exp |
               numb of int;
```

might be used to represent strictly bracketed integer arithmetic expressions:

```
<expression> ::= (<expression> + <expression>) |
                 (<expression> - <expression>) |
                 (<expression> * <expression>) |
                 (<expression> / <expression>) |
                 <integer>
```

so:

```
(<expression1> + <expression2>) ==
 add(<expression1>,<expression2>)
(<expression1> - <expression2>) ==
 diff(<expression1>,<expression2>)
(<expression1> * <expression2>) ==
 mult(<expression1>,<expression2>)
(<expression1> / <expression2>) ==
 quot(<expression1>,<expression2>)
<integer> == numb(<integer>)
```

For example:

```
1 == numb(1)
(1 + 2) == add(numb(1),numb(2))
((1 * 2) + 3) == add(mult(numb(1),numb(2)),numb(3))
((1 * 2) + (3 - 4)) == add(mult(numb(1),numb(2)),
                           diff(numb(3),numb(4)))
```

Write a function which evaluates an arithmetic expression in this representation, for example:

```
- eval (numb(1));
> 1 : exp

- eval (add(numb(1),numb(2)));
> 3 : exp

- eval (add(mult(numb(1),numb(2)),numb(3)));
> 5 : exp

- eval (add(mult(numb(1),numb(2)),diff(numb(3),numb(4))));
> 1 : exp
```

Chapter 10
Functional programming and LISP

LISP (LISt Programming) is a widely used artificial intelligence language. It is weakly typed with run-time type checking. Functions calls are evaluated in applicative order. LISP lacks structure and pattern matching.

Although LISP is not a pure functional language, it has a number of functional features. Like λ calculus, LISP has an incredibly simple core syntax. This is based on bracketed symbol sequences which may be interpreted as list data structures or as function calls. The shared representation of data and program reputedly makes LISP particularly appropriate for artificial intelligence applications.

The original LISP programming system was devised by McCarthy in the late 1950s as an aid to the Advice Taker experimental artificial intelligence system. McCarthy's early description of LISP was based on a functional formalism influenced by λ calculus, known as **Meta expressions (M-expressions)**. These were represented in an extremely simple **Symbolic expression (S-expression)** format for practical programming. Contemporary LISP systems are based solely on the S-expression format, although other functional languages are reminiscent of the richer M-expressions. We will not consider M-expressions here.

LISP, like BASIC, is not a unitary language and is available in a number of widely differing dialects. The heart of these differences, as we shall see, lies in the way that name/object associations are treated. Here, we will consider Common LISP which is a modern standard. We will also look briefly at Scheme, a fully functional LISP. Other LISPs include Franz LISP which is widely available on UNIX systems, MacLISP and INTERLISP which Common LISP subsumes, and Lispkit LISP which is another fully functional LISP.

It is important to note that LISP has been refined and developed for many years and so is not a very 'clean' language. LISP systems differ in how some aspects of LISP are implemented. Some aspects of LISP are extremely arcane and subject to much disputation amongst the cognoscenti. Furthermore, while LISP has much in common with functional languages, it is actually used as an imperative programming language for most applications.

We will only look at enough LISP to see how it corresponds to our functional approach. Many details will, necessarily, be omitted.

10.1 Atoms, numbers and symbols

The basic LISP objects are **atoms** composed of sequences of printing characters. Whenever a LISP system sees an atom, it tries to evaluate it.

Common LISP provides distinct representations for integer, ratio, floating point and complex **number** atoms. Here, we will only consider integers. These consist of digit sequences preceded by an optional sign, for example:

```
0  42  -99
```

The result of evaluating a number is that number.

Symbols or **literals** are non-numeric atoms and correspond to names, for example:

```
banana BANANA forty_two --> +
```

Symbols have associated values. The result of evaluating a symbol is its associated value. There are a large number of system symbols with standard associated values known as **primitives**. As we will see, symbols are also objects in their own right.

10.2 Forms, expressions and function applications

The **form** is the basic LISP construct and consists of an atom or a left bracket, followed by zero or more atoms or forms ending with a right bracket:

```
<form> ::= <atom> | ( <forms> ) | ()
<forms> ::= <form> | <form> <forms>
```

Forms are used for all expressions and data structures in LISP. Forms are always strictly bracketed except when special shorthand constructs are introduced.

Expressions are always prefix. The first form in a bracketed form sequence is interpreted as the **function**. Subsequent forms are interpreted as **arguments**. Thus, in a bracketed sequence of forms, we will refer to the first form as the function and to the subsequent forms as the arguments. We will also refer to primitives as if they were system functions.

Forms are evaluated in applicative order from left to right. For expressions, we will use -> to indicate a result after applicative order evaluation.

Note that the function form may be the name of a function or a lambda expression but may NOT be an expression returning a function!

Note also that the argument form may NOT be a lambda function or the name from a global definition or the name of a primitive!

Special primitives and techniques are used to treat functions as values.

10.3 Logic

The primitive for TRUE is:

```
t
```

and:

```
nil
```

is the primitive for FALSE.

```
not and or
```

are the primitives for the logical negation, conjunction and disjunction functions respectively. These may be used to construct simple logical expressions as forms, for example:

```
(not t) ->
nil

(and t nil) ->
nil

(or (and t nil) (not nil)) ->
t
```

In LISP, unlike most programming languages, and and or may have more than two arguments. For and, the final value is the conjunction of all the arguments, for example:

```
(and t nil t) ->
nil
```

For or, the final value is the disjunction of all the arguments, for example:

```
(or t nil t) ->
t
```

10.4 Arithmetic and numeric comparison

The primitives for the addition, subtraction, multiplication and division functions are:

```
+ - * /
```

These may be used with numbers to construct simple arithmetic expressions, for example:

```
(+ 40 2) ->
42

(- 46 4) ->
42

(* 6 (+ 3 4)) ->
42

(/ (+ 153 15) (- 7 3)) ->
42
```

As with and and or, these functions may have more than two arguments, so '+' returns the sum, '–' the difference, '*' the product and '/' the overall quotient, for example:

```
(+ 12 25 5) ->
42

(- 59 8 9) ->
42

(* 3 2 7) ->
42

(/ 336 4 2) ->
42
```

'/' is actually a real division operator.

The primitive:

```
truncate
```

rounds a single argument down to the nearest integer. If truncate is given two arguments, then it divides one by the other and rounds the result down to the nearest integer, for example:

```
(truncate 43 6) ->
7
```

The primitive:

```
rem
```

returns the integer remainder after division, for example:

```
(rem 43 6) ->
1
```

The numeric less than, less than or equal, equality, greater than or equal and greater than primitive comparison functions are:

```
< <= = >= >
```

These may all be used with more than two arguments. Thus '=' checks that all its arguments are equal, for example:

```
(= 2 2 2 2 2) ->
t
```

'<=' checks that its arguments are in ascending order, for example:

```
(<= 1 2 2 3 4 5) ->
t
```

'<' checks that its arguments are in strictly ascending order, for example:

```
(< 1 2 2 3 4 5) ->
nil
```

'>=' checks that its arguments are in descending order, for example:

```
(>= 5 4 3 3 2 1) ->
t
```

and '>' checks that its arguments are in strictly descending order, for example:

```
(> 9 8 7 6 5) ->
t
```

The primitive:

```
numberp
```

returns true if its argument is a number. For example,

```
(numberp 42) ->
t
```

10.5 Lambda functions

LISP uses a notation like that for λ calculus to define nameless functions. It is important to note that LISP functions do not have all the properties we might expect from λ calculus. In particular, special techniques are needed to pass functions as arguments, to return functions as values and to apply functions returned as values to new arguments.

Functions are defined as forms with the primitive:

```
lambda
```

followed by a flat list of bound variables and the body form:

```
(lambda (<bound variables>) <body>)
```

where:

```
<bound variables> ::= <bound variable> |
                      <bound variable> <bound variables>
```

For example, to square a number:

```
(lambda (x) (* x x))
```

or to find the sum of the squares of two numbers:

```
(lambda (x y) (+ (* x x) (* y y)))
```

or to find the value of the quadratic:

```
ax²+bx+c
```

given a, b, c and x:

```
(lambda (a b c x) (+ (* a (* x x)) (* b x) c)))
```

Note that functions are normally uncurried in LISP.

Note also that LISP systems will reject attempts to present λ functions directly as values other than in the function position in a form. lambda is not a primitive which denotes a system function. Instead, it acts as a marker to indicate a function form. However, if a LISP system sees a naked lambda function form, it will try to find a function associated with lambda and fail. The special techniques needed to manipulate function values are discussed below.

A function is applied to arguments in a form with the function followed by the arguments. The function's body is evaluated with the bound variables associated initially with the corresponding arguments:

```
(<function> <argument1> <argument2> ... )
```

Note that arguments for uncurried functions are not bracketed but follow straight after the function. For example:

```
((lambda (x) (* x x)) 2) ->
4

((lambda (x y) (+ (* x x) (* y y))) 3 4) ->
25

((lambda (a b c x) (+ (* a (* x x)) (* b x) c))) 1 2 1 1) ->
4
```

10.6 Global definitions

LISP systems are usually interactive and accept forms from the input for immediate evaluation. Global definitions provide a way of naming functions for use in subsequent forms.

A definition is a form with the primitive:

```
defun
```

followed by the name, bound variable list and the body form:

```
(defun <name> (<bound variable list>) <body>)
```

Many LISP systems print the defined name after a definition. For example:

```
(defun sq (x) (* x x)) ->
sq

(defun sum_sq (x y) (+ (* x x) (* y y))) ->
sum_sq

(defun quad (a b c x) (+ (* a (* x x)) (* b x) c)) ->
quad
```

A defined name may then be used instead of a λ function in other forms. For example:

```
(sq 2) ->
4

(sum_sq 3 4) ->
25

(quad 1 2 1 2) ->
9
```

In particular, defined names may be used in other definitions. For example, the last two definitions might be shortened to:

```
(defun sum_sq (x y) (+ (sq x) (sq y)))

(defun quad (a b c x) (+ (* a (sq x)) (* b x) c))
```

It is important to note that a global definition establishes a special sort of name/value relationship which is not the same as that between bound variables and values.

10.7 Conditional expressions

LISP conditional expressions are forms with the primitive:

```
cond
```

followed by a sequence of options. Each option is a test expression followed by a result expression which is to be evaluated if the test expression is true:

```
(cond (<test1> <result1>)
      (<test2> <result2>)
      ...
      (t <resultN>))
```

Note that the last option's test expression is usually t to ensure a final value for the conditional.

When a conditional expression is evaluated, each option's test expression is tried in turn. When a true test expression is found, the value of the corresponding result expression is returned. This is like a nested sequence of if ... then ... else expressions. Note that a conditional expression is not evaluated in applicative order. For example, to find the larger of two values:

```
(defun max (x y)
 (cond ((> x y) x)
       (t y)))
```

or to define logical not:

```
(defun lnot (x)
 (cond (x nil)
       (t t)))
```

or to define logical and:

```
(defun land (x y)
 (cond (x y)
       (t nil)))
```

Common LISP provides a simpler conditional primitive:

```
if
```

which is followed by a test and expressions to be evaluated if the test is true or false:

```
(if <test>
   <true result>
   <false result>)
```

For example, to find the smaller of two numbers:

```
(defun min (x y)
 (if (< x y)
   x
   y))
```

or to define logical or:

```
(defun lor (x y)
  (if x
     t
     y))
```

10.8 Quoting

We said above that, in LISP, forms are used as both program and data structures. So far, we have used forms as function calls in general and to build higher level control structures using special system primitives which control form evaluation. In order to use forms as data structures, additional primitives are used to prevent form evaluation. This is a different approach to λ calculus where data structures are packaged as functions with bound variables to control the subsequent application of selector functions.

In LISP, a mechanism known as **quoting** is used to delay form evaluation. This is based on the idea from ordinary language use, that if you want to refer to something's name, rather than that thing, then you put the name in quotation marks. For example:

```
Edinburgh is in Scotland
```

is a statement about a city, whereas:

```
'Edinburgh' has nine letters
```

is a statement about a word. Putting in the quotation marks shows that we are interested in the letter sequence rather than the thing to which the letter sequence refers. In LISP, quoting is used to prevent form evaluation. Quoting a form shows that we are interested in the sequence of subforms as a structure, rather than the form's final value. Quoting is a special sort of abstraction for delaying evaluation. Later on, we will see how it can be reversed.

When the LISP primitive:

```
quote
```

is applied to an argument, then that argument is returned unevaluated:

```
(quote <argument>) -> <argument>
```

In particular, a symbol argument is not replaced by its associated value but becomes an object in its own right. Quoting is so widely used in LISP that the special notation:

```
'<argument>
```

has been introduced. This is equivalent to the above use of the quote primitive.

10.9 Lists

LISP is perhaps best known for the use of list processing as the basis of programming.

The empty list is the primitive:

```
nil
```

which may be also written as:

```
()
```

The tester for an empty list is the primitive:

```
null
```

This is the same as not because FALSE is nil in LISP and anything which is not FALSE is actually TRUE!

Lists may be constructed explicitly with the:

```
cons
```

primitive:

```
(cons <head> <tail>)
```

The <head> and <tail> arguments are evaluated, and a list with value <head> in the head and value <tail> in the tail, is formed.

If the eventual tail is not the empty list, then the **dotted pair** representation is used for the resultant list:

```
<head value> . <tail value>
```

For example:

```
(cons 1 2) ->
1 . 2

(cons 1 (cons 2 3)) ->
1. (2 . 3)

(cons (cons 1 2) (cons 3 4)) ->
(1 . 2) . (3 . 4)

(cons (cons 1 2) (cons (cons 3 4) (cons 5 6))) ->
(1 . 2) . ((3 . 4) . (5 . 6))
```

When a list ends with the empty list, then a flat representation based on forms is used with an implicit empty list at the end. For example:

```
(cons 1 nil) ->
(1)

(cons 1 (cons 2 nil)) ->
(1 2)

(cons (cons 1 (cons 2 nil)) (cons (cons 3 (cons 4 nil)) nil)) ->
((1 2) (3 4))
```

Thus, as in Chapter 6, the empty list nil is equivalent to the empty form '( )'. Note that lists built by cons and ending with the empty list appear to be forms, but are not actually evaluated further as function calls.

Lists may be constructed directly in form notation and this is the most common approach. Note, however, that list forms must be explicitly quoted to prevent function call evaluation. For example:

```
(1 2 3)
```

looks like a call to the function 1 with arguments 2 and 3, whereas:

```
'(1 2 3)
```

is the list:

```
1 . (2 . (3 . nil))
```

The primitive:

```
list
```

is a multi-argument version of cons but constructs a list ending with the empty list. For example:

```
(list 1 2) ->
(1  2)

(list 1 2 3) ->
(1 2 3)

(list (list 1 2) (list 3 4)) ->
((1 2) (3 4))

(list (list 1 2) (list 3 4) (list 5 6)) ->
((1 2) (3 4) (5 6))
```

The primitive:

```
listp
```

returns true if its argument is a list. For example:

```
(listp '(1 2 3 4)) ->
t
```

10.10 List selection

The head of a LISP list is known as the **car** and the tail as the **cdr**. This is from the original IBM 704 implementation where a list head was processed as the 'Contents of the Address Register' and the tail as the 'Contents of the Decrement Register'. Thus:

```
car
```

is the head selection primitive and:

```
cdr
```

is the tail selection primitive. For example:

```
(car '(1 2 3)) ->
1

(cdr '(1 2 3)) ->
(2 3)

(car (cdr '(1 2 3))) ->
2

(cdr (cdr '(1 2 3))) ->
(3)
```

Note that sublists selected from lists appear to be forms, but are not further evaluated as function call forms by car or cdr.

10.11 Recursion

In LISP, recursive functions are based on function definitions with the defined name appearing in the function body. For example, to find the length of a linear list:

```
(defun length (l)
 (if (null l)
    0
    (+ 1 (length (cdr l)))))
```

and to count how often a value appears in a linear list of numbers:

```
(defun count (x l)
 (cond ((null l) 0)
        ((= x (car l)) (+ 1 (count x (cdr l))))
        (t (count x (cdr l)))))
```

Or, to insert a value into an ordered list:

```
(defun insert (x l)
 (cond ((null l) (cons x nil))
        ((< x (car l)) (cons x l))
        (t (cons (car l) (insert x (cdr l))))))
```

and to sort a list:

```
(defun sort (l)
 (if (null l)
    nil
    (t (insert (car l) (sort (cdr l))))))
```

10.12 Local definitions

LISP provides the:

```
let
```

primitive for the introduction of local variables in expressions:

```
(let ((<variable1> <value1>)
     (<variable2> <value2>)
     ...)
     (<result>))
```

which is equivalent to the function call:

```
((lambda (<variable1> <variable2> ... ) <result>)
 <value1> <value2> ... )
```

For example, to insert a value into an ordered list if it is not there already:

```
(defun new_insert (x l)
 (if (null l)
    (cons x nil)
    (let ((hl (car l))
          (tl (cdr l)))
         (cond ((= x hl) l)
               ((< x hl) (cons x l))
               (t (cons hl (new_insert x tl)))))))
```

Here, if the list is empty, then a new list with the value in the head is returned. Otherwise, the head and tail of the list are selected. If the head matches the value, then the list is returned. If the value comes before the head, then it is added before the head. Otherwise the value is inserted in the tail.

10.13 Binary trees in LISP

In Chapter 7, we looked at the construction of binary trees using a list representation. This translates directly into LISP: we will use nil for the empty tree and the list:

```
(<item> <left> <right>)
```

for the tree with node value <item>, left branch <left> and right branch <right>. We will use:

```
(defun node (item left right) (list item left right))
```

to construct new nodes.

LISP has no pattern matching so it is useful to define selector functions:

```
(defun item (l) (car l))

(defun left (l) (car (cdr l)))

(defun right (l) (car (cdr (cdr l))))
```

for the node value, left branch and right branch.

Thus, to add an integer to an ordered binary tree:

```
(defun tadd (i tree)
 (cond ((null tree) (node v nil nil))
       ((< i (item tree)) (node (item tree)
                                (tadd i (left tree)) (right tree)))
       (t (node (item tree) (left tree) (tadd i (right tree))))))
```

For example:

```
(tadd 7 nil) ->
(7 nil nil)

(tadd 4 '(7 nil nil)) ->
(7
 (4 nil nil)
 nil)

(tadd 10 '(7 (4 nil nil) nil)) ->
(7
 (4 nil nil)
 (10 nil nil))

(tadd 2 '(7 (4 nil nil) (10 nil nil))) ->
(7
 (4
  (2 nil nil)
  nil)
 (10 nil nil))

(tadd 5 '(7 (4 (2 nil nil) nil) (10 nil nil))) ->
(7
 (4
  (2 nil nil)
  (5 nil nil))
 (10 nil nil))
```

Hence, to add a list of numbers to an ordered binary tree:

```
(defun taddlist (l tree)
 (if (null l)
    tree
    (taddlist (cdr l) (tadd (car l) tree))))
```

Finally, to traverse a binary tree in ascending order:

```
(defun traverse (tree)
 (if (null tree)
    nil
    (append (traverse (left tree))
                  (cons (item tree) (traverse (right tree))))))
```

For example:

```
(traverse '(7 (4 (2 nil nil) (5 nil nil) (10 nil nil))) ->
(2 4 5 7 10)
```

10.14 Dynamic and lexical scope

LISP is often presented as if it were based on λ calculus but this is somewhat misleading. LISP function abstraction uses a *notation* similar to λ abstraction but the relationship between bound variables and variables in expressions is rather opaque.

In our presentation of λ calculus, names have **lexical** or **static** scope. That is, a name in an expression corresponds to the bound variable of the innermost enclosing function to define it. Consider the following contrived example. We might define:

```
def double_second = λx.λx.(x + x)
```

This is a function with bound variable:

```
x
```

and body:

```
λx.(x + x)
```

Thus, in the expression:

```
(x + x)
```

the xs correspond to the second rather than the first bound variable. We would normally avoid any confusion by renaming:

```
def double_second = λx.λy.(y + y)
```

For lexical scope, the bound variable corresponding to a name in an expression is determined by their relative positions in that expression, before the expression is evaluated. Expression evaluation cannot change that static relationship.

Early LISPs were based on **dynamic** scope where names' values are determined when expressions are evaluated. Thus, a name in an expression corresponds to the most recent bound variable/value association with the same name, when that name is encountered during expression evaluation. This is effectively the same as lexical scope when a name is evaluated in the scope of the corresponding statically scoped bound variable.

However, LISP functions may contain free variables and a function may be created in one scope and evaluated in another. Thus, a name might refer to different variables, depending on the scopes in which it is evaluated. Expression evaluation can change the name/bound variable correspondence. For example, suppose we want to calculate the tax on a gross income but do not know the rate of tax. Instead of making the tax rate a bound variable, we could make it a free variable. Our approach to λ calculus does not allow this, unless the free variable has been introduced by a previous definition. However, this is allowed in LISP:

```
(defun tax (gross)
 (/ (* gross rate) 100))
```

Note that rate is free in the function body. For a LISP with dynamic scope, like Franz LISP, rate's value is determined when:

```
(/ (* gross rate) 100)
```

is evaluated. For example, if the lowest rate of tax is 25% then we might define:

```
(defun low_tax (gross)
 (let ((rate 25))
      (tax gross)))
```

When low_tax is applied to an argument, rate is set to 25 and then tax is called. Thus, rate in tax is evaluated in the scope of the local variable rate in low_tax and will have the value 25.

In LISPs with dynamic scope this use of free variables is seen as a positive advantage because it delays decisions about name/value associations. Thus, the same function with free variables may be used to different effect in different places. For example, suppose the average tax rate is 30%. We might define:

```
(defun av_tax (gross)
 (let ((rate 30))
     (tax gross)))
```

Once again, the call to tax in av_tax evaluates the free variable rate in tax in the scope of the local rate in av_tax, this time with value 30.

It is not clear whether or not dynamic scope was a conscious feature or the result of early approaches to implementing LISP. Common LISP is based on lexical scope, but provides a primitive to make dynamic scope explicit if it is needed. Attempts to move lexically scoped free variables in and out of different scopes are faulted in Common LISP.

10.15 Functions as values and arguments

In looking at λ calculus, we have become used to treating functions as objects which can be manipulated freely. This approach to objects is actually quite uncommon in programming languages, in part because until comparatively recently it was thought that it was hard to implement. In LISP, functions are not like other objects and cannot be simply passed around as in λ calculus. Instead, function values must be identified explicitly and applied explicitly, except in the special case of the function form in a function application form.

The provision of functions as **first class citizens** in LISP used to be known as the **FUNARG problem** because implementation problems centred on the use of functions with free variables as arguments to other functions. This arose because of dynamic scope where a free variable is associated with a value in the calling, rather than the defining, scope. However, it is often necessary to return a function value with free variables frozen to values from the defining scope. We have used this in defining typed functions in Chapter 5.

The traditional way round the FUNARG problem is to identify function values explicitly, so that free variables can be frozen in their defining scopes. The application of such function values is also made explicit, so that free variables are not evaluated in the calling scope. This freezing of free variables in a function value is often implemented by constructing a **closure** which identifies the relationship between free and lexical bound variables. Common LISP is based on lexical scope where names are frozen in their defining scope. None the less, Common LISP still requires function values to be identified and applied explicitly.

In Common LISP, the primitive:

```
function
```

is used as a special form of quoting for functions:

```
(function <function>)
```

It creates a function value in which free variables are associated with bound variables in the defining scope. As with quote, an equivalent special notation:

```
#'<function>
```

is provided. Note that most LISP systems will not actually display function values as text. This is because they translate functions into intermediate forms to ease implementation but lose the equivalent text. Some systems may display an implementation dependent representation.

For example, we could define a general tax function:

```
(defun gen_tax (rate)
 #'(lambda (gross) (/ (* gross rate) 100)))
```

We might then produce the low and average tax rate functions as:

```
(gen_tax 25)
```

and:

```
(gen_tax 30)
```

which return lambda function values with rate bound to 25 and 30 respectively.

The primitive:

```
funcall
```

is used to apply a function value to its arguments:

```
(funcall <function value> <argument1> <argument2> ...)
```

For example, to apply the low tax function to a gross income:

```
(funcall (gen_tax 25) 10000) ->
2500
```

This call builds a function value with rate bound to 25 and then applies it to the gross income 10000. Similarly:

```
(funcall (gen_tax 30) 15000) ->
4500
```

builds a function value with rate bound to 30 and then applies it to the gross income 15000.

The mapping function mapcar may be defined as:

```
(defun mapcar (fn arg)
 (if (null arg)
   nil
   (cons (funcall fn (car arg)) (mapcar fn (cdr arg)))))
```

Note that the function fn is applied explicitly to the argument car arg by funcall. Now, we could square every element of a list with:

```
(defun sq_list (l)
 (mapcar #'(lambda (x) (* x x)) l))
```

Note that the function argument for mapcar has been quoted with #'. Note also, that even if a function value argument is identified simply by name, then that name must still be quoted with #' before it may be used as an argument. For example, to apply sq to every element of a list:

```
(mapcar #'sq '(1 2 3 4 5))
```

10.16 Symbols, quoting and evaluation

Normally, symbols are variable names, but they may also be used as objects in their own right. Quoting a symbol prevents the associated value being extracted. Thereafter, a quoted symbol may be passed around like any object. One simple use for quoted symbols is as Pascal-like user-defined enumeration types. For example, we might define the days of the week as:

```
'Monday 'Tuesday 'Wednesday 'Thursday 'Friday 'Saturday 'Sunday
```

We could then write functions to manipulate these objects. In particular, they may be compared with the equality primitive which is true if its arguments are identical objects:

```
eq
```

Note that in Pascal, enumeration types are mapped onto integers and so have successor functions defined automatically for them. In LISP, we have to define a successor function explicitly if we need one. For example:

```
(defun next_day (day)
 (cond ((eq day 'Monday)  'Tuesday)
 ((eq day 'Tuesday)       'Wednesday)
 ((eq day 'Wednesday)     'Thursday)
 ((eq day 'Thursday)      'Friday)
 ((eq day 'Friday)        'Saturday)
 ((eq day 'Saturday)      'Sunday)
 ((eq day 'Sunday)        'Monday)))
```

The primitive:

```
eval
```

forces evaluation of a quoted form as a LISP expression. Thus, functions can construct quoted forms which are program structures for later evaluation by other functions. Thus, compiling techniques can be used to produce LISP from what is apparently data. For example, the rules for an expert system might be translated into functions to implement that system. Similarly, the grammar for an interface language might be used to generate functions to recognize that language. For example, suppose we want to translate strictly bracketed infix arithmetic expressions:

```
<expression> ::= <number> |
                 (<expression> + <expression>) |
                 (<expression> - <expression>) |
                 (<expression> * <expression>) |
                 (<expression> / <expression>)
```

into prefix form so:

```
(<expression> + <expression>) ==
 (+ <expression> <expression>)
(<expression> - <expression>) ==
 (- <expression> <expression>)
(<expression> * <expression>) ==
 (* <expression> <expression>)
(<expression> / <expression>) ==
 (/ <expression> <expression>)
```

We need to extract the operator and place it at the head of a list with the translated expressions as arguments:

```
(defun trans (l)
 (if (numberp l)
  l
  (let ((e1 (trans (car l)))
```

```
            (op (car (cdr l)))
            (e2 (trans (car (cdr (cdr l))))))
      (list op e1 e2))))
```

For example:

```
(trans '(1 + 2)) ->
(+ 1 2)

(trans '(3 * (4 * 5))) ->
(* 3 (* 4 5))

(trans '((6 * 7) + (8 - 9))) ->
(+ (* 6 7) (- 8 9))
```

Note that quoted symbols for infix operators have been moved into the function positions in the resultant forms. Now, we can evaluate the translated expression, as a LISP form, using eval:

```
(defun calc (l)
 (eval (trans l)))
```

For example:

```
(calc '((6 * 7) + (8 - 9))) ->
41
```

The treatment of free variables in quoted forms depends on the scope rules. For dynamic scope systems, quoted free variables are associated with the corresponding bound variable when the quoted form is evaluated. Thus, with dynamic scope, quoting can move free variables into different evaluation scopes.

In Common LISP, with lexical scope, quoted free variables are not associated with bound variables in the defining scope as might be expected. Instead, they are evaluated as if there were no bound variables defined, apart from the names from global definitions.

10.17 λ calculus in LISP

We can use function and funcall for rather clumsy applicative order pure λ calculus. For example, we can try out some of the λ functions we met in Chapter 2. First of all, we can apply the identity function:

```
#'(lambda (x) x)
```

to itself:

```
(funcall #'(lambda (x) x) #'(lambda (x) x))
```

Note that some LISP implementations will not print out the resultant function and others will print a system specific representation.

Let us now use definitions:

```
(defun identity (x) x) ->
identity

(funcall #'identity #'identity)
```

Note once again that there may be no resultant function or an internal representation will be printed.

Let us apply the self-application function to the identity function:

```
(funcall #'(lambda (s) (funcall s s)) #'(lambda (x) x))
```

Notice that we must make the application of the argument to itself explicit. Now, let us use definitions:

```
(defun self_apply (s) (funcall s s)) ->
self_apply

(funcall #'self_apply #'identity)
```

Finally, let us try the function application function:

```
(defun apply (f a) (funcall f a)) ->
apply
```

with the self-application function and the identity function:

```
(funcall #'apply #'self_apply #'identity)
```

Pure λ calculus in LISP is complicated by this need to make function values and their applications explicit and by the absence of uniform representations for function values.

10.18 λ calculus and Scheme

Scheme is a language in the LISP tradition which provides function values without explicit function identification and application. Scheme, like other LISPs, uses the bracket-based form as the combined program and data structure, is weakly typed and has applicative order evaluation. Like Common LISP, Scheme is lexically scoped. We are not going to look at Scheme in any depth. Here, we are only going to consider the use of function values.

In Scheme, functions are true first class objects. They may be passed as arguments, and function expressions may appear in the function position in forms. Thus, many of our pure λ calculus examples will run with a little translation into Scheme. However, Scheme systems may not display directly function values as text, but may produce a system dependent representation.

Let us consider once again the examples from Chapter 2. We can enter a λ function directly, for example, the identity function:

```
(lambda (x) x)
```

We can also directly apply functions to functions, for example, we might apply the identity function to itself:

```
((lambda (x) x) (lambda (x) x))
```

Alas, the representation of the result depends on the system.

Scheme function definitions have the form:

```
(define (<name> <argument1> <argument2> ...) <body>)
```

Defined functions may be applied without quoting:

```
(define (identity x) x) ->
identity

(identity identity)
```

Note once again that there may be no resultant function or a system specific internal representation may be printed.

To continue with the self-application and function application functions:

```
(lambda (s) (s s))

((lambda (s) (s s)) (lambda (x) x))

(define (self_apply s) (s s)) ->
self_apply

(self_apply identity)

(define (apply f a) (f a)) ->
apply

(apply self_apply identity)
```

This explicit use of function values is much closer to λ calculus than Common LISP, though there is still no uniform representation for function values.

10.19 Other features

It is impossible to consider an entire language in a small chapter. Here, we have concentrated on Common LISP in relation to pure functional programming. Common LISP has come a long way since the original LISP systems. In particular, it includes a wide variety of data types which we have not considered, including characters, arrays, strings, structures and multiple value objects. We also have not looked at input/output and other system interface facilities.

SUMMARY

- The relationship between Common LISP and our approach to functional programming has been examined.
- Algorithms from preceding chapters may be implemented in Common LISP.
- Treating functions as objects in Common LISP involves explicit notation; there is no standard representation for function results.
- Scheme simplifies treating functions as objects but lacks a standard representation for function results.

EXERCISES

10.1 See Exercises 4.1–4.4. Write functions to:

(a) Find the sum of the integers between n and 0.

(b) Find the product of the numbers between n and 1.

(c) Find the sum of applying a function fun to the numbers between n and 0.

(d) Find the sum of applying a function fun to the numbers between n and zero in steps of s.

10.2 See Exercise 6.2. Write a function which:

(a) Indicates whether or not a list starts with a sublist. For example:

```
(lstarts '(1 2 3) '(1 2 3 4 5)) ->
t

(lstarts '(1 2 3) '(4 5 6)) ->
nil
```

(b) Indicates whether or not a list contains a given sublist. For example:

```
(lcontains '(4 5 6) '(1 2 3 4 5 6 7 8 9)) ->
t
(lcontains '(4 5 6) '(2 4 6 8 10)) ->
nil
```

(c) Counts how often a sublist appears in another list. For example:

```
(lcount '(1 2) '(1 2 3 1 2 3 1 2 3)) ->
3
```

(d) Removes a sublist from the start of a list, assuming that you know that the sublist starts the list. For example:

```
(lremove '(1 2 3) '(1 2 3 4 5 6 7 8 9)) ->
(4 5 6 7 8 9))
```

(e) Deletes the first occurrence of a sublist in another list. For example:

```
(ldelete '(4 5 6) '(1 2 3 4 5 6 7 8 9) ->
(1 2 3 7 8 9)
(ldelete '(4 5 6) '(2 4 6 8 10)) ->
(2 4 6 8 10)
```

(f) Inserts a sublist after the first occurrence of another sublist in a list. For example:

```
(linsert '(4 5 6) '(1 2 3) '(1 2 3 7 8 9)) ->
(1 2 3 4 5 6 7 8 9)
(linsert '(4 5 6) '(1 2 3) '(2 4 6 8 10)) ->
(2 4 6 8 10)
```

(g) Replaces a sublist with another sublist in a list. For example:

```
(lreplace '(6 5 4) '(4 5 6) '(9 8 7 6 5 4 3 2 1)) ->
(9 8 7 4 5 6 3 2 1)
(lreplace '(6 5 4) '(4 5 6) '(2 4 6 8 10)) ->
(2 4 6 8 10)
```

10.3 See Exercise 6.3.

(a) Write a function which merges two ordered lists to produce an ordered list.

(b) Write a function which merges a list of ordered lists.

10.4 See Exercise 7.1. The time of day might be represented as a list with three integer fields for hours, minutes and seconds:

```
(<hours> <minutes> <seconds>)
```

For example:

```
(17 35 42) == 17 hours 35 minutes 42 seconds
```

Note that:

```
24 hours = 0 hours
1 hour == 60 minutes
1 minute == 60 seconds
```

(a) Write functions to convert from a time of day to seconds and from seconds to a time of day. For example:

```
(too_secs (2 30 25)) ->
9025

(from_secs 48975) ->
(13 36 15)
```

(b) Write a function which increments the time of day by one second. For example:

```
(tick (15 27 18)) ->
(15 27 19)

(tick (15 44 59)) ->
(15 45 0)

(tick (15 59 59) ->
(16 0 0)

(tick (23 59 59) ->
(0 0 0)
```

(c) In a shop, each transaction at a cash register is time stamped. Given a list of transaction details, where each is a string followed by a time of day, write a function which sorts them into ascending time order. For example:

```
(tsort '((apples (12 19 57))
         (tomatoes (18 22 48))
         (orange_juice (10 12 35))
```

```
(coffee (15 47 19)))) ->

(orange_juice (10 12 35))
(apples (12 19 57))
(coffee (15 47 19))
(tomatoes (18 22 48)))
```

10.5 See Exercise 7.2.

(a) Write a function which compares two integer binary trees.

(b) Write a function which indicates whether or not one integer binary tree contains another as a subtree.

(c) Write a function which traverses a binary tree to produce a list of node values in descending order.

Answers to exercises

Chapter 2

2.1 (a) λa.(a λb.(b a))

<function>
<bound variable> – a
<body> – (a λb.(b a))
<application>
<function exp> – <name> – a
<argument exp> – λb.(b a)
<function>
<bound variable> – b
<body> – (b a)
<application>
<function exp> – <name> – b
<argument exp> – <name> – a

(b) λx.λy.λz.((z x) (z y))

<function>
<bound variable> – x
<body> λy.λz.((z x) (z y))
<function>
<bound variable> – y
<body> – λz.((z x) (z y))
<function>
<bound variable> – z
<body> – ((z x) (z y))
<application>
<function exp> – (z x)
<application>
<function exp> – <name> – z
<argument exp> – <name> – x
<function exp> – (z y)
<application>
<function exp> – <name> – z
<argument exp> – <name> – y

(c) (λf.λg.(λh.(g h) f) λp.λq.p)

<application>
 <function exp> – λf.λg.(λh.(g h) f)
  <function>
   <bound variable> – f
   <body> – λg.(λh.(g h) f)
    <function>
     <bound variable> – g
     <body> – (λh.(g h) f)
      <application>
       <function exp> – λh.(g h)
        <function>
         <bound variable> – h
         <body> – (g h)
          <application>
           <function exp> – <name> – g
           <argument exp> – <name> – h
       <argument exp> – <name> f
 <argument exp> – λp.λq.p
  <function>
   <bound variable> – p
   <body> – λq.p
    <function>
     <bound variable> – q
     <body> – <name> – p

(d) λfee.λfi.λfo.λfum.(fum (fo (fi fee)))

<function>
 <bound variable> – fee
 <body> – λfi.λfo.λfum.(fum (fo (fi fee)))
  <function>
   <bound variable> – fi
   <body> – λfo.λfum.(fum (fo (fi fee)))
    <function>
     <bound variable> – fo
     <body> – λfum.(fum (fo (fi fee)))
      <function>
       <bound variable> – fum
       <body> – (fum (fo (fi fee)))
        <application>
         <function exp> – <name> – fum
         <argument exp> – (fo (fi fee))
          <application>
           <function exp> – <name> – fo
           <argument exp> – (fi fee)
            <application>
             <function exp> – <name> – fi
             <argument exp> – <name> – fee

(e) ((λp.(λq.p λx.(x p)) λi.λj.(j i)) λa.λb.(a (a b)))

```
<application>
 <function exp> - (λp.(λq.p λx.(x p)) λi.λj.(j i))
  <application>
   <function exp> - λp.(λq.p λx.(x p))
    <function>
     <bound variable> - p
     <body> - (λq.p λx.(x p))
      <application>
       <function exp> - λq.p
        <function>
         <bound variable> - q
         <body> - <name> - p
       <argument exp> - λx.(x p)
        <function>
         <bound variable> - x
         <body> - (x p)
          <application>
           <function exp> - <name> - x
           <argument exp> - <name> - p
   <argument exp> - λi.λj.(j i)
    <function>
     <bound variable> - i
     <body> - λj.(j i)
      <function>
       <bound variable> - j
       <body> - (j i)
        <application>
         <function exp> - <name> - j
         <argument exp> - <name> - i
 <argument exp> - λa.λb.(a (a b))
  <function>
   <bound variable> - a
   <body> - λb.(a (a b))
    <function>
     <bound variable> - b
     <body> - (a (a b))
      <application>
       <function exp> - <name> - a
       <argument exp> - (a b)
        <application>
         <function exp> - <name> - a
         <argument exp> - <name> - b
```

2.2 (a) ((λx.λy.(y x) λp.λq.p) λi.i) =>
(λy.(y λp.λq.p) λi.i) =>
(λi.i λp.λq.p) =>
λp.λq.p

(b)
```
(((λx.λy.λz.((x y) z) λf.λa.(f a)) λi.i) λj.j) =>
((λy.λz.((λf.λa.(f a) y) z) λi.i) λj.j) =>
(λz.((λf.λa.(f a) λi.i) z) λj.j) =>
((λf.λa.(f a) λi.i) λj.j) =>
(λa.(λi.i a) λj.j) =>
(λi.i λj.j) =>
λj.j
```

(c)
```
(λh.((λa.λf.(f a) h) h) λf.(f f)) =>
((λa.λf.(f a) λf.(f f)) λf.(f f)) =>
(λf.(f λf.(f f)) λf.(f f)) =>
(λf.(f f) λf.(f f)) =>
(λf.(f f) λf.(f f)) => ...
```

(d)
```
((λp.λq.(p q) (λx.x λa.λb.a)) λk.k) =>
(λq.((λx.x λa.λb.a) q) λk.k) =>
((λx.x λa.λb.a) λk.k) =>
(λa.λb.a λk.k) =>
λb.λk.k
```

(e)
```
(((λf.λg.λx.(f (g x)) λs.(s s)) λa.λb.b) λx.λy.x) =>
((λg.λx.(λs.(s s) (g x)) λa.λb.b) λx.λy.x) =>
(λx.(λs.(s s) (λa.λb.b x)) λx.λy.x) =>
(λs.(s s) (λa.λb.b λx.λy.x)) =>
((λa.λb.b λx.λy.x) (λa.λb.b λx.λy.x)) =>
(λb.b (λa.λb.b λx.λy.x)) =>
(λa.λb.b λx.λy.x) =>
λb.b
```

2.3 (a) (i)
```
(identity <argument>) => ... =>
<argument>
```

(ii)
```
((apply (apply identity)) <argument>) => ... =>
((apply identity) <argument>) => ... =>
(identity <argument>) => ... =>
<argument>
```

(b) (i)
```
((apply <function>) <argument>) => ... =>
(<function> <argument>)
```

(ii)
```
((λx.λy.(((make_pair x) y) identity) <function>)
  <argument>) =>
(λy.(((make_pair <function>) y) identity) <argument>) =>
(((make_pair <function>) <argument>) identity) => ... =>
((identity <function>) <argument>) => ... =>
(<function> <argument>)
```

(c) (i)
```
(identity <argument>) => ... =>
<argument>
```

```
(ii)((self_apply (self_apply select_second))
     <argument>) => ... =>
    (((self_apply select_second) (self_apply select_second))
     <argument>) => ... =>
    (((select_second select_second) (self_apply select_second))
     <argument>) => ... =>
    ((λsecond.second (self_apply select_second)
     <argument>) => ... =>
    ((select_second select_second) <argument>) => ... =>
    (λsecond.second <argument>) =>
    <argument>
```

2.4

```
def make_triplet = λfirst.
                    λsecond.
                     λthird.
                      λs.(((s first) second) third)

def triplet_first = λfirst.λsecond.λthird.first

def triplet_second = λfirst.λsecond.λthird.second

def triplet_third = λfirst.λsecond.λthird.third

make_triplet <item1> <item2> <item3> triplet_first ==
λfirst.
 λsecond.
  λthird.
   λs.(((s first) second) third)
<item1> <item2> <item3> triplet_first => ... =>
(((triplet_first <item1>) <item2>) <item3>) ==
(((λfirst.λsecond.λthird.first <item1>) <item2>) <item3>) => ... =>
<item1>

make_triplet <item1> <item2> <item3> triplet_first ==
λfirst.
 λsecond.
  λthird.
   λs.(((s first) second) third)
<item1> <item2> <item3> triplet_second => ... =>
(((triplet_second <item1>) <item2>) <item3>) ==
(((λfirst.λsecond.λthird.second <item1>) <item2>)
 <item3>) => ... =>
<item2>

make_triplet <item1> <item2> <item3> triplet_third ==
λfirst.
 λsecond.
  λthird.
   λs.( (s first) second) third)
<item1> <item2> <item3> triplet_third => ... =>
```

```
(((triplet_third <item1>) <item2>) <item3>) ==
(((λ first.λsecond.λthird.third <item1>) <item2>) <item3>)
 => ... =>
<item3>
```

2.5 (a)

```
      λx.λy.(λx.y λy.x)
   x bound at {x} in λx.λy.(λx.y λy.{x})
   x free at {x} in λy.(λx.y λy.{x})
                    (λx.y λy.{x})
                    λy.{x}
                    {x}
   y bound at {y} in λx.λy.(λx.{y} λy.x)
                      λy.(λx.{y} λy.x)
   y free at {y} in (λx.{y} λy.x)
                    λx.{y}
                    {y}
```

(b)

```
      λx.(x (λy.(λx.x y) x))
   x bound at {x} in λx.({x} (λy.(λx.x y) {x}))
   x free at {x} in ({x} (λy.(λx.x y) {x}))
                 in {x}
                 in λy.(λx.x y) {x})
                 in {x}
   x bound at {x} in λy.(λx.{x} y)
                      (λx.{x} y)
                      λx.{x}
   x free at {x} in {x}
   y bound at {y} in λx.(x (λy.(λx.x {y}) x))
                      (x (λy.(λx.x {y}) x))
                      (λy.(λx.x {y}) x)
                      λy.(λx.x {y})
   y free at {y} in (λx.x {y})
                    {y}
```

(c)

```
      λa.(λb.a λb.(λa.a b))
   a bound at {a} in λa.(λb.{a} λb.(λa.a b))
   a free at {a} in (λb.{a} λb.(λa.a b))
                    λb.{a}
                    {a}
   a bound at {a} in λa.(λb.a λb.(λa.{a} b))
                      (λb.a λb.(λa.{a} b))
                      λb.(λa.{a} b))
                      λa.{a}
   a free at {a} in {a}
   b bound at {b} in λa.(λb.a λb.(λa.a {b}))
                      (λb.a λb.(λa.a {b}))
                      λb.(λa.a {b})
   b free at {b} in (λa.a {b})
                    {b}
```

```
(d) (λfree.bound λbound.(λfree.free bound))
    bound free at {bound} in
     (λfree.{bound} λbound.(λfree.free bound))
     λfree.{bound}
     {bound}
    bound bound at {bound} in
     (λfree.bound λbound.(λfree.free {bound}))
     λbound.(λfree.free {bound})
    bound free at {bound} in (λfree.free {bound})
                              {bound}
    free bound at {free} in
     (λfree.bound λbound.(λfree.{free} bound))
     λbound.(λfree.{free} bound)
     (λfree.{free} bound)
     λfree.{free}
    free free at {free} in {free}

(e) λp.λq.(λr.(p (λq.(λp.(r q)))) (q p))
    p bound at {p} in λp.λq.(λr.({p} (λq.(λp.(r q)))) (q {p}))
    p free at {p} in λq.(λr.({p} (λq.(λp.(r q)))) (q {p}))
                     (λr.({p} (λq.(λp.(r q)))) (q {p}))
                     λr.({p} (λq.(λp.(r q))))
                     ({p} (λq.(λp.(r q))))
                     {p}
                     (q {p})
                     {p}
    q bound at {q} in λp.λq.(λr.(p (λq.(λp.(r q)))) ({q} p))
                      λq.(λr.(p (λq.(λp.(r q)))) ({q} p))
    q free at {q} in (λr.(p (λq.(λp.(r q)))) ({q} p))
                     ({q} p))
                     {q}
    q bound at {q} in λp.λq.(λr.(p (λq.(λp.(r {q})))) (q p))
                      λq.(λr.(p (λq.(λp.(r {q})))) (q p))
                      (λr.(p (λq.(λp.(r {q})))) (q p))
                      λr.(p (λq.(λp.(r {q}))))
                      (p (λq.(λp.(r {q}))))
                      λq.(λp.(r {q}))
    q free at {q} in (λp.(r {q}))
                     (r {q})
                     {q}
    r bound at {r} in λp.λq.(λr.(p (λq.(λp.({r} q)))) (q p))
                      λq.(λr.(p (λq.(λp.({r} q)))) (q p))
                      (λr.(p (λq.(λp.({r} q)))) (q p))
                      λr.(p (λq.(λp.({r} q))))
    r free at {r} in (p (λq.(λp.({r} q))))
                     λq.(λp.({r} q)))
                     (λp.({r} q))
                     ({r} q)
                     {r}
```

2.6 (a) λx.λy.(λz.y λa.x)

(b) λx.(x (λy.(λz.z y) x))

(c) λa.(λb.a λb.(λc.c b))

(e) λp.λq.(λr.(p (λs.(λt.(r s)))) (q p))

Chapter 3

3.1 def implies = λx.λy.(x y true)

implies false false => ... => false false true => ... => true
implies false true => ... => false true true => ... => true
implies true false => ... => true false true => ... => false
implies true true => ... => true true true => ... => true

3.2 def equiv = λx.λy.(x y (not y))

equiv false false => ... => false false (not false) => ... => true
equiv false true => ... => false true (not true) => ... => false
equiv true false => ... => true false (not false) => ... => false
equiv true true => ... => true true (not true) => ... => true

3.3 (a) (i) λx.λy.(and (not x) (not y)) false false => ... =>
and (not false) (not false) => ... =>
(not false) (not false) false) => ... =>
true (not false) false => ... =>
not false => ... => true

λx.λy.(and (not x) (not y)) false true => ... =>
and (not false) (not true) => ... =>
(not false) (not true) false => ... =>
true (not true) false => ... =>
not true => ... => false

λx.λy.(and (not x) (not y)) true false => ... =>
and (not true) (not false) => ... =>
(not true) (not false) false => ... =>
false (not false) false => ... => false

λx.λy.(and (not x) (not y)) true true => ... =>
and (not true) (not true)
(not true) (not true) false => ... =>
false (not true) false => ... => false

(ii)λx.λy.(not (or x y)) false false => ... =>
not (or false false) => ... =>
(or false false) false true => ... =>
(false true false) false true => ... =>
false false true => ... => true

λx.λy.(not (or x y)) false true => ... =>
not (or false true) => ... =>
(or false true) false true => ... =>
(false true true) false true => ... =>
true false true => ... => false

λx.λy.(not (or x y)) true false => ... =>
not (or true false) => ... =>
(or true false) false true => ... =>
(true true false) false true => ... =>
true false true => ... => false

λx.λy.(not (or x y)) true true => ... =>
not (or true true) => ... =>
(or true true) false true => ... =>
(true true true) false true => ... =>
true false true => ... => false

(b) (i) See 3.1 above.

(ii)λx.λy.(implies (not y) (not x)) false false => ... =>
implies (not false) (not false) => ... =>
(not false) (not false) true => ... =>
true (not false) true => ... =>
not false => ... => true

λx.λy.(implies (not y) (not x)) false true => ... =>
implies (not true) (not false) => ... =>
(not true) (not false) true => ... =>
false (not false) true => ... => true

λx.λy.(implies (not y) (not x)) true false => ... =>
implies (not false) (not true)
(not false) (not true) true => ... =>
true (not true) true => ... =>
not true => ... => false

λx.λy.(implies (not y) (not x)) true true => ... =>
implies (not true) (not true) => ... =>
(not true) (not true) true => ... =>
false (not true) true => ... => true

(c) (i) not false => ... => true
not true => ... => false

(ii)λx.(not (not (not x))) false =>
not (not (not false)) => ... =>
(not (not false)) false true => ... =>
((not false) false true) false true => ... =>
((false false true) false true) false true => ... =>
(true false true) false true => ... =>
false false true => ... => true
λx.(not (not (not x))) true =>
not (not (not true)) => ... =>
((not (not true)) false true) => ... =>
((not true) false true) false true => ... =>
((true false true) false true) false true => ... =>
(false false true) false true => ... =>
true false true => ... => false

(d) (i) See 3.1 above.

(ii)λx.λy.(not (and x (not y))) false false => ... =>
not (and false (not false)) => ... =>
(and false (not false)) false true => ... =>
(false (not false) false) false true) => ... =>
false false true => ... => true

λx.λy.(not (and x (not y))) false true => ... =>
not (and false (not true)) => ... =>
(and false (not true)) false true => ... =>
(false (not true) false) false true => ... =>
false false true => ... => true

λx.λy.(not (and x (not y))) true false => ... =>
not (and true (not false)) => ... =>
(and true (not false)) false true => ... =>
(true (not false) false) false true => ... =>
(not false) false true => ... =>
true false true => ... => false

λx.λy.(not (and x (not y))) true true => ... =>
not (and true (not true)) => ... =>
(and true (not true)) false true => ... =>
(true (not true) false) false true => ... =>
(not true) false true => ... =>
false false true => ... => true

(e) (i) See 3.2 above.

```
(ii)λx.λy.(and (implies x y) (implies y x)) false false => ... =>
   and (implies false false) (implies false false) => ... =>
   (implies false false) (implies false false) false => ... =>
   (false false true) (implies false false) false => ... =>
   true (implies false false) false => ... =>
   implies false false => ... =>
   false false true => ... => true

   λx.λy.(and (implies x y) (implies y x)) false true => ... =>
   and (implies false true) (implies true false)
   (implies false true) (implies true false) false => ... =>
   (false true true) (implies true false) false => ... =>
   true (implies true false) false => ... =>
   implies true false => ... =>
   true false true => ... => false

   λx.λy.(and (implies x y) (implies y x)) true false => ... =>
   and (implies true false) (implies false true)
   (implies true false) (implies false true) false => ... =>
   (true false true) (implies false true) false => ... =>
   false (implies false true) false => ... => false

   λx.λy.(and (implies x y) (implies y x)) true true => ... =>
   and (implies true true) (implies true true) => ... =>
   (implies true true) (implies true true) false => ... =>
   (true true true) (implies true true) false => ... =>
   true (implies true true) false => ... =>
   implies true true => ... =>
   true true true => ... => true
```

3.4

```
λx.(succ (pred x)) λs.(s false <number>) =>
succ (pred λs.(s false <number>)
Simplifying: pred λs.(s false <number>) => ... =>
             <number>
so: succ <number> => ... =>
    λs.(s false <number>)

λx.(pred (succ x)) λs.(s false <number>) =>
pred (succ λs.(s false number>))
Simplifying: succ λs.(s false <number>) => ... =>
             λs.(s false λs.(s false <number>))
so: pred λs.(s false λs.(s false <number>)) => ... =>
    λs.(s false <number>)

λx.(succ (pred x)) zero =>
(succ (pred zero))
Simplifying: pred zero => ... => zero
```

```
so: succ zero ==
    one

λx.(pred (succ x)) zero =>
(pred (succ zero))
Simplifying: succ zero == one
so: pred one => ... =>
    zero
```

Chapter 4

4.1

```
sum three => ... =>
recursive sum1 three => ... =>
sum1 (recursive sum1) three => ... =>
add three ((recursive sum1) (pred three)) -> ... ->
add three (sum1 (recursive sum1) two) -> ... ->
add three (add two ((recursive sum1) (pred two))) -> ... ->
add three (add two (sum1 (recursive sum1) one)) -> ... ->
add three (add two (add one ((recursive sum1) (pred one))))
  -> ... ->
add three (add two (add one (sum1 (recursive sum1) zero)))
  -> ... ->
add three (add two (add one zero)) -> ... ->
six
```

4.2

```
def prod1 f n =
 if equal n one
 then one
 else mult n (f (pred n))

def prod = recursive prod1

prod three
recursive prod1 three => ... =>
prod1 (recursive prod1) three => ... =>
mult three ((recursive prod1) (pred three)) -> ... ->
mult three (prod1 (recursive prod1) two) -> ... ->
mult three (mult two ((recursive prod1) (pred two))) -> ... ->
mult three (mult two (prod1 (recursive prod1) one)) -> ... ->
mult three (mult two one) -> ... ->
six
```

4.3

```
def fun_sum1 f fun n =
 if iszero n
 then fun zero
 else add (fun n) (f fun (pred n))
```

```
def fun_sum = recursive fun_sum1

fun_sum double three => ... =>
recursive fun_sum1 double three => ... =>
fun_sum1 (recursive fun_sum1) double three => ... =>
add (double three)
    ((recursive fun_sum1) double (pred three)) -> ... ->
add (double three)
    (fun_sum1 (recursive fun_sum1) double two) -> ... ->
add (double three)
    (add (double two)
         ((recursive fun_sum1) double (pred two))) -> ... ->
add (double three)
    (add (double two)
         (fun_sum1 (recursive fun_sum1) double one)) -> ... ->
add (double three)
    (add (double two)
         (add (double one)
              ((recursive fun_sum1) double (pred one))))
 -> ... ->
add (double three)
    (add (double two)
         (add (double one)
              (fun_sum1 (recursive fun_sum1) double zero)))
 -> ... ->
add (double three)
    (add (double two)
         (add (double one)
              (double zero))) -> ... ->
twelve
```

4.4

```
def fun_sum_step1 f fun n s =
 if iszero n
 then fun n
 else add (fun n) (f fun (sub n s) s)
def fun_sum_step = recursive fun_sum_step1
```

(a)

```
fun_sum_step double five two => ... =>
recursive fun_sum_step1 double five two => ... =>
fun_sum_step1 (recursive fun_sum_step1) double five two
 => ... =>
add (double five)
    ((recursive fun_sum_step1) double (sub five two) two)
 -> ... ->
```

```
add (double five)
    (fun_sum_step1 (recursive fun_sum_step1)
     double three two) -> ... ->
add (double five)
    (add (double three)
     ((recursive fun_sum_step1) double (sub three two) two))
 -> ... ->
add (double five)
    (add (double three)
         (fun_sum_step1 (recursive fun_sum_step1)
          double one two)) -> ... ->
add (double five)
    (add (double three)
         (add (double one)
              ((recursive fun_sum_step1)
               double (sub one two) two))) -> ... ->
add (double five)
    (add (double three)
         (add (double one)
              (fun_sum_step1 (recursive fun_sum_step1)
               double zero two))) -> ... ->
add (double five)
    (add (double three)
         (add (double one)
              (double zero))) -> ... ->
eighteen
```

(b)
```
fun_sum_step double four two
recursive fun_sum_step1 double four two => ... =>
fun_sum_step1 (recursive fun_sum_step1) double four two
 => ... =>
add (double four)
    ((recursive fun_sum_step1) double (sub four two) two)
 -> ... ->
add (double four)
    (fun_sum_step1 (recursive fun_sum_step1) double two two)
  -> ... ->
add (double four)
    (add (double three)
     ((recursive fun_sum_step1) double (sub two two) two))
 -> ... ->
add (double four)
    (add (double two)
         (fun_sum_step1 (recursive fun_sum_step1)
                              double zero two)) -> ... ->
```

```
add (double four)
    (add (double two)
         (double zero))
twelve
```

4.5

```
def less x y = greater y x

def less_or_equal x y = greater_or_equal y x
```

(a)
```
less three two => ... =>
greater two three => ... =>
not (iszero (sub two three)) -> ... ->
not (iszero zero) -> ... ->
not true => ... => false
```

(b)
```
less two three => ... =>
greater three two -> ... -> true - see 4.7.3
```

(c)
```
less two two => ... =>
greater two two => ... =>
not (iszero (sub two two)) -> ... ->
not (iszero zero) -> ... ->
not true => ... => false
```

(d)
```
less_or_equal three two => ... =>
greater_or_equal two three => ... =>
iszero (sub three two) -> ... ->
iszero one => ... => false
```

(e)
```
less_or_equal two three => ... =>
greater_or_equal three two => ... =>
iszero (sub two three) -> ... ->
iszero zero => ... => true
```

(f)
```
less_or_equal two two => ... =>
greater_or_equal two two => ... =>
iszero (sub two two) -> ... ->
iszero zero => ... => true
```

4.6

```
def mod x y =
 if iszero y
 then x
 else mod1 x y
```

```
rec mod1 x y =
 if less x y
 then x
 else mod1 (sub x y) y
```

(a)
```
mod three two => ... =>
mod1 three two
mod1 (sub three two) two -> ... ->
mod1 one two => ... => one
```

(b)
```
mod two three => ... =>
mod1 two three => ... => two
```

(c)
```
mod three zero => ... => three
```

Chapter 5

5.1 (a)
```
ISBOOL 3 => ... =>
MAKE_BOOL (isbool 3) ==
MAKE_BOOL (istype bool_type 3) -> ... ->
MAKE_BOOL (equal (type 3) bool_type) -> ... ->
MAKE_BOOL (equal numb_type bool_type) -> ... ->
MAKE_BOOL false ==
FALSE
```

(b)
```
ISNUMB FALSE => ... =>
MAKE_BOOL (isnumb FALSE) ==
MAKE_BOOL (istype numb_type FALSE) -> ... ->
MAKE_BOOL (equal (type FALSE) numb_type) -> ... ->
MAKE_BOOL (equal bool_type numb_type) -> ... ->
MAKE_BOOL false ==
FALSE
```

(c)
```
NOT 1 => ... =>
if isbool 1
then MAKE_BOOL (not (value 1))
else BOOL_ERROR -> ... ->
if equal (type 1) bool_type
then ...
else ... -> ... ->
if equal numb_type bool_type
then ...
else ... -> ... ->
if false
then ...
else BOOL_ERROR => ... =>
BOOL_ERROR
```

(d)
```
TRUE AND 2 => ... =>
if and (isbool TRUE) (isbool 2)
then MAKE_BOOL (and (value TRUE) (value 2))
else BOOL_ERROR -> ... ->
if and (istype bool_type TRUE) (istype bool_type 2)
then ...
else ... -> ... ->
if and (equal (type TRUE) bool_type) (equal (type 2) bool_type)
then ...
else ... -> ... ->
if and (equal bool_type bool_type) (equal numb_type bool_type)
then ...
else ... -> ... ->
if and true false
then ...
else ... -> ... ->
if false
then ...
else BOOL_ERROR => ... =>
BOOL_ERROR
```

(e)
```
2 + TRUE => ... =>
if and (isnumb 2) (isnumb TRUE)
then MAKE_NUMB (add (value 2) (value TRUE))
else NUMB_ERROR -> ... ->
if and (istype numb_type 2) (istype numbtype TRUE)
then ...
else ... -> ... ->
if and (equal (type 2) numb_type)
        (equal (type TRUE) numb_type)
then ...
else ... -> ... ->
if and (equal numb_type numb_type)
        (equal numb_type numb_type)
then ...
else ... -> ... ->
if and true false
then ...
else NUMB_ERROR => ... =>
NUMB_ERROR
```

5.2 (a)
```
def issigned N = istype signed_type N

def ISSIGNED N = MAKE_BOOL (issigned N)

def sign = value (select_first (value N))

def SIGN N =
 if issigned N
 then select_first (value N)
 else SIGN_ERROR
```

```
    def sign_value N = value (select_second (value N))

    def VALUE N =
     if issigned N
     then select_second (value N)
     else SIGN_ERROR

    def sign_iszero N = iszero (sign_value N)

(b) def SIGN_ISZERO N =
     if issigned N
     then MAKE_BOOL (sign_iszero N)
     else SIGN_ERROR

    def SIGN_SUCC N =
     IF SIGN_ISZERO N
     THEN +1
     ELSE
      IF SIGN N
      THEN MAKE_SIGNED
            POS (MAKE_NUMB (succ (sign_value N)))
      ELSE MAKE_SIGNED
            NEG (MAKE_NUMB (pred (sign_value N)))

    def SIGN_PRED N =
     IF SIGN_ISZERO N
     THEN -1
     ELSE
      IF SIGN N
      THEN MAKE_SIGNED
              POS (MAKE_NUMB (pred (sign_value N)))
      ELSE MAKE_SIGNED
              NEG (MAKE_NUMB (succ (sign_value N)))

(c) def SIGN_+ X Y =
     if and (issigned X) (issigned Y)
     then
      if iszero (sign_value X)
      then Y
      else
       if sign_iszero (sign_value Y)
       then X
       else
        if and (sign X) (sign Y)
        then MAKE_SIGNED POS (MAKE_NUMB
               (add (sign_value X) (sign_value Y)))
        else
         if and (not (sign X)) (not (sign Y))
         then MAKE_SIGNED NEG (MAKE_NUMB
                (add (sign_value X) (sign_value Y)))
         else
```

```
    if not (sign X)
    then
     if greater (sign_value X) (sign_value Y)
     then MAKE_SIGNED NEG (MAKE_NUMB
              (sub (sign_value X) (sign_value Y)))
     else MAKE_SIGNED POS (MAKE_NUMB
              (sub (sign_value Y) (sign_value X)))
    else
     if greater (sign_value Y) (sign_value X)
     then MAKE_SIGNED NEG (MAKE_NUMB
              (sub (sign_value Y) (sign_value X)))
     else MAKE_SIGNED POS (MAKE_NUMB
              (sub (sign_value X) (sign_value Y)))
else SIGN_ERROR
```

Chapter 6

6.1

```
def ATOMCONS A L =
 if isnil L
 then [A]
 else
  if equal (type A) (type (HEAD L))
  then CONS A L
  else LIST_ERROR
```

6.2

(a)

```
rec STARTS [] L = TRUE
 or STARTS L [] = FALSE
 or STARTS (H1::T1) (H2::T2) =
  IF CHAR_EQUALS H1 H2
  THEN STARTS T1 T2
  ELSE FALSE
```

(b)

```
rec CONTAINS L [] = FALSE
 or CONTAINS L1 L2 =
  IF STARTS L1 L2
  THEN TRUE
  ELSE CONTAINS L1 (TAIL L2)
```

(c)

```
rec COUNT L [] = 0
 or COUNT L1 L2 =
  IF STARTS L1 L2
  THEN 1 + (COUNT L1 (TAIL L2))
  ELSE COUNT L1 (TAIL L2)
```

(d)

```
rec REMOVE [] L = L
 or REMOVE (H1::T1) (H2::T2) = REMOVE T1 T2
```

(e)
```
rec DELETE L [] = []
 or DELETE L1 L2 =
  IF STARTS L1 L2
  THEN REMOVE L1 L2
  ELSE (HEAD L2)::(DELETE L1 (TAIL L2))
```

(f)
```
rec INSERT L1 L2 [] = []
 or INSERT L1 L2 L3 =
  IF STARTS L2 L3
  THEN APPEND L2 (APPEND L1 (REMOVE L2 L3))
  ELSE (HEAD L3)::(INSERT L1 L2 (TAIL L3))
```

(g)
```
rec REPLACE L1 L2 [] = []
 or REPLACE L1 L2 L3 =
  IF STARTS L2 L3
  THEN APPEND L1 (REMOVE L2 L3)
  ELSE (HEAD L3)::(REPLACE L1 L2 (TAIL L3))
```

6.3 (a)
```
rec MERGE L [] = L
 or MERGE [] L = L
 or MERGE (H1::T1) (H2::T2) =
  IF LESS H1 H2
  THEN H1::(MERGE T1 (H2::T2))
  ELSE H2::(MERGE (H1::T1) T2)
```

(b)
```
rec LMERGE [] = []
 or LMERGE (H::T) = MERGE H (LMERGE T)
```

Chapter 7

7.1 (a)
```
def TOO_SECS [H,M,S] = (60 * ((60 * H) + M)) + S

def MOD X Y = X - ((X / Y) * Y)

def FROM_SECS S =
 let SECS = MOD S 60
 in
  let MINS = (MOD S 3600) / 60
  in
   let HOURS = S / 3600
   in [HOURS,MINS,SECS]
```

(b)
```
def TICK [H,M,S] =
    let S = S + 1
    in
     IF LESS S 60
     THEN [H,M,S]
```

```
    ELSE
     let M = M + 1
     in
      IF LESS M 60
      THEN [H,M,0]
      ELSE
       let H = H + 1
       in
        IF LESS H 24
        THEN [H,0,0]
        ELSE [0,0,0]
```

(c)
```
def TLESS [TR1,[H1,M1,S1]] [TR2,[H2,M2,S2]] =
 IF LESS H1 H2
 THEN TRUE
 ELSE
  IF EQUAL H1 H2
  THEN
   IF LESS M1 M2
   THEN TRUE
   ELSE
    IF EQUAL M1 M2
    THEN
     IF LESS S1 S2
     THEN TRUE
     ELSE FALSE
   ELSE FALSE
 ELSE FALSE

    rec TINSERT T [ ] = [T]
     or TINSERT T (T1::R) =
       IF TLESS T T1
       THEN T::T1::R
       ELSE T1::(TINSERT T R)

    rec TSORT [ ] = [ ]
     or TSORT (H::T) = TINSERT H (TSORT T)
```

7.2 (a)
```
rec TCOMP TEMPTY TEMPTY = TRUE
 or TCOMP TEMPTY T = FALSE
 or TCOMP T TEMPTY = FALSE
 or TCOMP [V1,L1,R1] [V2,L2,R2] =
  IF EQUAL V1 V2
  THEN AND (TCOMP L1 L2) (TCOMP R1 R2)
  ELSE FALSE
```

(b)
```
rec TFIND TEMPTY T = TRUE
 or TFIND T TEMPTY = FALSE
 or TFIND T1 T2 =
  IF TCOMP T1 T2
  THEN TRUE
  ELSE
   IF LESS (ITEM T1) (ITEM T2)
   THEN TFIND T1 (LEFT T2)
   ELSE TFIND T1 (RIGHT T2)
```

(c)
```
rec DTRAVERSE TEMPTY = []
 or DTRAVERSE [V,L,R] = APPEND (DTRAVERSE R)
                               (V::(DTRAVERSE L))
```

7.3
```
rec EVAL [E1,OP,E2] =
 let R1 = EVAL E1
 in
  let R2 = EVAL E2
  in
   IF STRING_EQUAL OP "+"
   THEN R1 + R2
   ELSE
    IF STRING_EQUAL OP "-"
    THEN R1 - R2
    ELSE
     IF STRING_EQUAL OP "*"
     THEN R1 * R2
     ELSE R1 / R2
or EVAL N = N
```

Chapter 8

8. 1 (a) Normal order

```
λs.(s s) (λf.λa.(f a) λx.x λy.y) =>
(λf.λa.(f a) λx.x λy.y) (λf.λa.(f a) λx.x λy.y) =>
(λa.(λx.x a) λy.y) (λf.λa.(f a) λx.x λy.y) =>
(λx.x λy.y) (λf.λa.(f a) λx.x λy.y) =>
λy.y (λf.λa.(f a) λx.x λy.y) =>
λf.λa.(f a) λx.x λy.y =>
λa.(λx.x a) λy.y =>
λx.x λy.y =>
λy.y
```

8 reductions

λf.λa.(f a) λx.x λy.y reduced twice

Applicative order
λs.(s s) (λf.λa.(f a) λx.x λy.y) ->
λs.(s s) (λa.(λx.x a) λy.y) ->
λs.(s s) (λx.x λy.y) ->
λs.(s s) λy.y ->
λy.y λy.y ->
λy.y

5 reductions
λf.λa.(f a) λx.x λy.y reduced once

Lazy
λs.(s s) (λf.λa.(f a) λx.x λy.y)$_1$ =>
(λf.λa.(f a) λx.x λy.y)$_1$ (λf.λa.(f a) λx.x λy.y)$_1$ =>
(λa.(λx.x a) λy.y)$_2$ (λa.(λx.x a) λy.y)$_2$ =>
(λx.x λy.y)$_3$ (λx.x λy.y)$_3$ =>
λy.y λy.y =>
λy.y

5 reductions
λf.λa.(f a) λx.x λy.y reduced once

(b) Normal order
λx.λy.x λx.x (λs.(s s) λs.(s s)) =>
λy.λx.x (λs.(s s) λs.(s s)) =>
λx.x

2 reductions
λs.(s s) λs.(s s) not reduced

Applicative order
λx.λy.x λx.x (λs.(s s) λs.(s s)) ->
λx.λy.x λx.x (λs.(s s) λs.(s s)) -> ...

Non-terminating – 1 reduction/cycle
λs.(s s) λs.(s s) reduced every cycle

Lazy
λx.λy.x λx.x (λs.(s s) λs.(s s)) =>
λy.λx.x (λs.(s s) λs.(s s)) =>
λx.x

2 reductions – as normal order

(c) Normal order
λa.(a a) (λf.λs.(f (s s)) λx.x) =>
(λf.λs.(f (s s)) λx.x) (λf.λs.(f (s s)) λx.x) =>
λs.(λx.x (s s)) (λf.λs.(f (s s)) λx.x) =>
λx.x ((λf.λs.(f (s s)) λx.x) (λf.λs.(f (s s)) λx.x)) =>
(λf.λs.(f (s s)) λx.x) (λf.λs.(f (s s)) λx.x) => ...

Non-terminating – 3 reductions/cycle
λf.λs.(f (s s)) λx.x reduced every cycle

Applicative order
```
λa.(a a) (λf.λs.(f (s s)) λx.x) ->
λa.(a a) λs.(λx.x (s s)) ->
λs.(λx.x (s s)) λs.(λx.x (s s)) ->
λx.x (λs.(λx.x (s s)) λs.(λx.x (s s))) ->
λs.(λx.x (s s)) λs.(λx.x (s s)) -> ...
```

Non-terminating – 2 reductions/cycle
λf.λs.(f (s s)) λx.x reduced before non-terminating cycle

Lazy

$\lambda a.(a\ a)\ (\lambda f.\lambda s.(f\ (s\ s))\ \lambda x.x)_1$ =>
$(\lambda f.\lambda s.(f\ (s\ s))\ \lambda x.x)_1\ (\lambda f.\lambda s.(f\ (s\ s))\ \lambda x.x)_1$ =>
λs.(λx.x (s s)) λs.(λx.x (s s)) =>
λx.x (λs.(λx.x (s s)) λs.(λx.x (s s))) =>
λs.(λx.x (s s)) λs.(λx.x (s s)) => ...

Non-terminating – 2 reductions/cycle
λf.λs.(f (s s)) λx.x reduced before non-terminating cycle

Chapter 9

9.1 (a)
```
fun cube (y:int) = y*y*y;
```

(b)
```
fun implies (x:bool) (y:bool) = (not x) orelse y;
```

(c)
```
fun smallest (a:int) (b:int) (c:int) =
 if a<b
 then
  if a<c
  then a
  else c
 else
  if b<c
  then b
  else c;
```

(d)
```
fun desc_join (s1:string) (s2:string) =
 if s1<s2
 then s1^s2
 else s2^s1;
```

(e)
```
fun shorter (s1:string) (s2:string) =
 if (size s1) < (size s2)
 then s1
 else s2;
```

9.2 (a)
```
fun sum 0 = 0 |
    sum (n:int) = n+(sum (n-1));
```

(b)
```
fun nsum (m:int) (n:int) =
  if m>n
  then 0
  else m+(nsum (m+1) n);
```

(c)
```
fun repeat (s:string) 0 = "" |
    repeat (s:string) (n:int) = s^(repeat s (n-1));
```

9.3 (a)
```
fun ncount [] = 0 |
    ncount ((h::t):int list) =
     if h < 0
     then 1+(ncount t)
     else ncount t;
```

(b)
```
fun scount (s:string) [] = 0 |
    scount (s:string) ((h::t):string list) =
     if h = s
     then 1+(scount s t)
     else scount s t;
```

(c)
```
fun gconst (v:int) [] = [] |
    gconst (v:int) ((h::t):int list) =
     if h > v
     then h::(gconst v t)
     else gconst v t;
```

(d)
```
fun smerge [] (s2:string list) = s2 |
    smerge (s1:string list) [] = s1 |
    smerge ((h1::t1):string list) ((h2::t2):string list) =
     if h1 < h2
     then h1::(smerge t1 (h2::t2))
     else h2::(smerge (h1::t1) t2);
```

(e)
```
fun slmerge [] = [] |
    slmerge ((h::t):(string list) list) = smerge h (slmerge t);
```

(f) (i)
```
type stock = string * int * int;
fun item (s:string,n:int,r:int) = s;
fun numb (s:string,n:int,r:int) = n;
fun reord (s:string,n:int,r:int) = r;

fun getmore [] = [] |
    getmore ((h::t):stock list) =
     if (numb h) < (reord h)
     then h::(getmore t)
     else getmore t;
```

```
(ii) type upd = string * int;
     fun uitem (s:string,n:int) = s;
     fun unumb (s:string,n:int) = n;

     fun update1 [] (u:upd) = [] |
         update1 ((h::t):stock list) (u:upd) =
          if (item h) = (uitem u)
          then (item h,(numb h)+(unumb u),reord h)::t
          else h::(update1 t u);

    fun update (r:stock list) [] = r |
        update (r:stock list) ((h::t):upd list) =
         update (update1 r h) t;
```

9.4 (a)
```
fun left1 0 (s:string list) = "" |
    left1 (n:int) [] = "" |
    left1 (n:int) ((h::t):string list) = h^(left1 (n-1) t);

fun left (n:int) (s:string) = left1 n (explode s);
```

(b)
```
fun drop 0 (s:string list) = s |
    drop (n:int) [] = [] |
    drop (n:int) ((h::t):string list) = drop (n-1) t;

fun right (n:int) (s:string) = implode (drop ((size s)-n)
                                              (explode s));
```

(c)
```
fun middle (n:int) (l:int) (s:string) = left1 l (drop (n-1)
                                                  (explode s));
```

(d)
```
fun starts [] (s2:string list) = true |
    starts (s1:string list) [] = false |
    starts ((h1::t1):string list) ((h2::t2):string list) =
     if h1=h2
     then starts t1 t2
     else false;

fun find1 [] (s2:string list) = 1 |
    find1 (s1:string list) [] = 1 |
    find1 (s1:string list) (s2:string list) =
     if starts s1 s2
     then 1
     else 1+(find1 s1 (tl s2));

fun find (s1:string) (s2:string) =
 let val pos = find1 (explode s1) (explode s2)
 in
```

```
    if pos > (size s2)
    then 0
    else pos
end;
```

9.5

```
fun east Queen_Street = Bishopbriggs |
    east Bishopbriggs = Lenzie |
    east Lenzie = Croy |
    east Croy = Polmont |
    east Polmont = Falkirk_High |
    east Falkirk_High = Linlithgow |
    east Linlithgow = Haymarket |
    east Haymarket = Waverly |
    east Waverly = Waverly;

fun west Queen_Street = Queen_Street |
    west Bishopbriggs = Queen_Street |
    west Lenzie = Bishopbriggs |
    west Croy = Lenzie |
    west Polmont = Croy |
    west Falkirk_High = Polmont |
    west Linlithgow = Falkirk_High |
    west Haymarket = Linlithgow |
    west Waverly = Haymarket;
```

9.6

```
fun eval (numb(i:int)) = i |
    eval (add(e1:exp,e2:exp)) = (eval e1)+(eval e2) |
    eval (diff(e1:exp,e2:exp)) = (eval e1)-(eval e2) |
    eval (mult(e1:exp,e2:exp)) = (eval e1)*(eval e2) |
    eval (quot(e1:exp,e2:exp)) = (eval e1) div (eval e2);
```

Chapter 10

10.1 (a)

```
(defun nsum (n)
  (if (eq 0 n)
      0
      (+ n (nsum (- n 1)))))
```

(b)

```
(defun nprod (n)
  (if (eq 1 n)
      1
      (* n (nprod (- n 1)))))
```

(c)

```
(defun napply (fun n)
  (if (eq 0 n)
```

```
        (funcall fun 0)
        (+ (funcall fun n) (napply fun (- n 1)))))
```

(d)
```
(defun nstepapply (fun n s)
    (if (<= n 0)
        (funcall fun 0)
        (+ (funcall fun n) (nstepapply fun (- n s) s))))
```

10.2 (a)
```
(defun lstarts (l1 l2)
    (cond ((null l1) t)
          ((null l2) nil)
          ((eq (car l1) (car l2)) (lstarts (cdr l1) (cdr l2)))
          (t nil)))
```

(b)
```
(defun lcontains (l1 l2)
    (cond ((null l2) nil)
          ((lstarts l1 l2) t)
          (t (lcontains l1 (cdr l2)))))
```

(c)
```
(defun lcount (l1 l2)
    (cond ((null l2) 0)
          ((lstarts l1 l2) (+ 1 (lcount l1 (cdr l2))))
          (t (lcount l1 (cdr l2)))))
```

(d)
```
(defun lremove (l1 l2)
    (if (null l1)
        l2
        (lremove (cdr l1) (cdr l2))))
```

(e)
```
(defun ldelete (l1 l2)
    (cond ((null l2) nil)
          ((lstarts l1 l2) (lremove l1 l2))
          (t (cons (car l2) (ldelete l1 (cdr l2))))))
```

(f)
```
(defun linsert (l1 l2 l3)
    (cond ((null l3) nil)
          ((lstarts l2 l3) (append l2 (append l1 (lremove l2 l3))))
          (t (cons (car l3) (linsert l1 l2 (cdr l3))))))
```

(g)
```
(defun lreplace (l1 l2 l3)
    (cond ((null l3) nil)
```

```
            ((lstarts l1 l3) (append l2 (lremove l1 l3)))
            (t (cons (car l3) (lreplace l1 l2 (cdr l3))))))
```

10.3 (a)
```
(defun merge (l1 l2)
  (cond ((null l1) l2)
        ((null l2) l1)
        ((< (car l1) (car l2)) (cons (car l1) (merge (cdr l1) l2)))
        (t (cons (car l2) (merge l1 (cdr l2))))))
```

(b)
```
(defun lmerge (l)
  (if (null l)
    nil
    (merge (car l) (lmerge (cdr l)))))
```

10.4 (a)
```
(defun hours (hms) (car hms))

(defun mins (hms) (car (cdr hms)))

(defun secs (hms) (car (cdr (cdr hms))))

(defun too_secs (hms)
 (+ (* 60 (+ (* 60 (hours hms)) (mins hms))) (secs hms)))

(defun from_secs (s)
 (list (truncate s 3600)
       (truncate (rem s 3600) 60)
       (rem s 60)))
```

(b)
```
(defun tick (hms)
  (let ((h (hours hms))
        (m (mins hms))
        (s (secs hms)))
        (let ((s1 (+ s 1)))
             (if (< s1 60)
                (list h m s1)
                (let ((m1 (+ m 1)))
                     (if (< m1 60)
                        (list h m1 0)
                        (let ((h1 (+ h 1)))
                             (if (< h1 24)
                                (list h1 0 0)
                                (list 0 0 0)))))))))
```

(c)
```
(defun hms (trans) (car (cdr trans)))
```

```
(defun tless (tr1 tr2)
 (let ((t1 (hms tr1))
      (t2 (hms tr2)))
      (let ((h1 (hours t1))
            (m1 (mins t1))
            (s1 (secs t1))
            (h2 (hours t2))
            (m2 (mins t2))
            (s2 (secs t2)))
           (if (< h1 h2)
              t
              (if (= h1 h2)
                 (if (< m1 m2)
                    t
                    (if (= m1 m2)
                       (if (< s1 s2)
                          t
                          nil)
                       nil))
                 nil)))))
```

```
(defun tinsert (tr l)
 (cond ((null l) (cons tr nil))
       ((tless tr (car l)) (cons tr l))
       (t (cons (car l) (tinsert tr (cdr l))))))
```

```
(defun tsort (l)
 (if (null l)
    l
    (tinsert (car l) (tsort (cdr l)))))
```

10.5 (a)
```
(defun tcomp (t1 t2)
 (cond ((and (null t1) (null t2)) t)
       ((or (null t1) (null t2)) nil)
       ((= (item t1) (item t2)) (and (tcomp (left t1) (left t2))
                                     (tcomp (right t1) (right t2))))
       (t nil)))
```

(b)
```
(defun tfind (t1 t2)
 (cond ((null t1) t)
       ((null t2) nil)
       ((tcomp t1 t2) t)
       ((< (item t1) (item t2)) (tfind t1 (left t2)))
       (t (tfind t1 (right t2)))))
```

(c)
```
(defun dtraverse (tree)
  (if (null tree)
    nil
    (append (dtraverse (right tree))
            (cons (item tree) (dtraverse (left tree))))))
```

Bibliography

This is an introductory book which draws on a wide range of sources. Most of the material here is covered in other texts though often with different perspectives and emphases. This bibliography is not exhaustive: rather, it offers a broad cross-section of source and supplementary material. References are almost entirely to books on the grounds that these tend to be more discursive than academic papers.

Chapter 1 Introduction

The influential paper by Backus (1978) contains a critique of von Neumann computing and arguments for functional programming. Brief motivational material on functional programming is contained in Sadler and Eisenbach (1987), Glaser *et al.* (1984), Henderson (1980) and Henson (1987).

Brady (1977) provides an accessible introduction to the theory of computing. There are further accounts in a wide variety of mathematical logic texts, of which Kleene's (1952) and Mendelson's (1964) are still outstanding.

For related general computing topics not considered further in this book: the denotational approach to programming language semantics is covered by Gordon (1979), Schmidt (1986) and Stoy (1977); program specification is covered by Cohen *et al.* (1986), Gehani and McGettrick (1986), Hayes (1987), Jones (1986) and more formally by Turski and Maibaum (1987); program verification is covered by Backhouse (1986), Gries (1981), Manna (1974) and less formally though thoroughly by Bornat (1987).

For related functional programming and language topics not considered further in this book: Glaser *et al.* (1984) contains overviews of implementation techniques; SECD machine implementations are discussed by Brady (1977), Burge (1975), Field and Harrison (1988), Henderson (1980), Henson (1987) and Wegner (1971); combinators and graph reduction are covered thoroughly by Field and Harrison (1988) and by Peyton-Jones (1987) and, in less detail, by Boutel (1987), Cripps *et al.* (1987), Hankin *et al.* (1987), Henson (1987) and Revesz (1988); Bird and Wadler (1988) cover verification; Field and Harrison (1988) cover program transformation and abstract interpretation; Henson (1987) covers verification and transformation; Harrison and Khoshnevisan (1987) also discuss transformation.

For functional languages not considered further in this book: Field and Harrison (1988) is based on and Hope and also contains brief discussion of

Miranda, LISP and FP; Peyton-Jones (1987) is based on Miranda and contains an introduction by Turner (1987); Bailey (1987) covers Hope; Harrison and Khoshnevisan (1987) cover FP; Henson (1987) covers FP; Glaser *et al.* (1984) discuss briefly FP, Hope and KRC; Bird and Wadler (1988) contains a brief appendix on Miranda; Revesz (1988) discusses briefly FP and Miranda; SASL is covered by Turner (1976).

For various imperative languages mentioned here: ALGOL 60, Naur *et al.* (1963); ALGOL 68, Pagan (1976); BASIC, Cope (1981); BCPL, Richards and Whitby-Strevens (1982); C, Kernighan and Ritchie (1978); Pascal, Findlay and Watt (1981); POP-2, Burstall *et al.* (1977); PROLOG, Clocksin and Mellish (1981) and PS-ALGOL, Carrick *et al.* (1986).

Chapter 2 λ calculus

The description of λ calculus by Church (1941) is much referenced but little read. Barendregt (1981) is the standard reference. Hindley and Seldin (1987) is less detailed.

Early descriptions from a computing perspective include Burge (1975) and Wegner (1971). Field and Harrison (1988), Peyton-Jones (1987) and Revesz (1988) provide thorough contemporary accounts, as does Stoy (1977) though oriented to semantics. Glaser *et al.* (1984) and Henson (1987) also describe λ calculus.

Pair functions are discussed by Barendregt (1981), Field and Harrison (1988), Glaser *et al.* (1984), Henson (1987), Revesz (1988) and Wegner (1971).

Chapter 3 Conditions, booleans and numbers

Burge (1975), Field and Harrison (1988), Glaser *et al.* (1984), Henson (1987), Revesz (1988) and Wegner (1971) all have accounts of aspects of the material on conditional expressions and booleans. Schmidt (1986) and Stoy (1977) cover it in exercises.

There are many approaches to representing numbers. The approach here is discussed by Barendregt (1981), Glaser *et al.* (1984) and also by Revesz (1988) in an exercise. Field and Harrison (1988) and Wegner (1971) discuss variants. Church's representation is discussed by Barendregt (1981), Burge (1975), Henson (1987), Revesz (1988) and Wegner (1971).

Chapter 4 Recursion and arithmetic

Barendregt (1981), Brady (1977), Field and Harrison (1988), Glaser *et al.* (1984), Henson (1987), Peyton-Jones (1987), Revesz (1988) and Stoy (1977) all discuss the derivation of the 'recursion' function. Schmidt (1986) presents it in an exercise. Burge (1975) discusses it in terms of combinators.

Kleene (1952), Mendelson (1964), Peter (1967) and Rayward-Smith (1986) provide accounts of the construction of arithmetic and comparison operations within recursive function theory. Brief computing oriented accounts are in Burge (1975) and Glaser *et al.* (1984).

Chapter 5 Types

The approach to types considered here is a λ calculus implementation of run-time typing with tags. Abelson and Sussman (1985) discuss a related Scheme approach to manifest types, and Henderson (1980) and Peyton-Jones (1987) discuss implementations of typing.

Field and Harrison (1988) and Peyton-Jones (1987) contain thorough accounts of polymorphism and type checking. Strachey (1967) and Cardelli (1983) discuss polymorphism.

Chapter 6 Lists and strings

Bird and Wadler (1988), Henderson (1980), Henson (1987) and Revesz (1988) contain thorough accounts of lists from a functional programming perspective.

Numerous books on LISP contain material on list processing and mapping functions. Wilensky (1986) provides an accessible introduction.

Chapter 7 Composite values and trees

The use of lists to represent composite values is discussed implicitly in numerous books on LISP.

Henderson (1980) and Queinnec (1983) discuss accumulation variables. Abelson and Sussman (1985) and Shapiro (1986) discuss the list representation of binary trees in Scheme and LISP respectively.

Bird and Wadler (1988) provides a thorough account of trees from a functional language perspective.

Chapter 8 Evaluation

Barendregt (1981) and Hindley and Seldin (1987) contain formal details of reduction. There are more accessible approaches in Brady (1977), Burge (1975), Field and Harrison (1988), Glaser *et al.* (1984), Henson (1987), Peyton-Jones (1987), Revesz (1988), Stoy (1977) and Wegner (1971). Note that Barendregt, Burge, Field and Harrison, Hindley and Seldin, and Wegner only name one Church-Rosser theorem.

Bird and Wadler (1988) discusses evaluation models, time and space efficiency and algorithm design techniques.

Brady (1977) and Minsky (1972) contain details of the halting problem for Turing machines.

Abelson and Sussman (1985), Bird and Wadler (1988), Field and Harrison (1988), Glaser *et al.* (1984), Henderson (1980), Henson (1987), Peyton-Jones (1987), and Revesz (1988) all discuss aspects of lazy evaluation.

Bird and Wadler (1988) provides a thorough account of infinite lists.

Chapter 9 Functional programming in Standard ML

Wikstrom (1987) provides thorough coverage of SML programming. Harper *et al.* (1986) describes SML informally and gives details of I/O and modules. (The informal description and the I/O details are duplicated in Wikstrom.) Harper *et al.* (1987) is a first version of a formal semantics for SML.

Bird and Wadler (1988) provides much additional material which is relevant to SML programming including discussion of concrete and abstract types.

Chapter 10 Functional programming and LISP

Steele (1984) is the Common LISP 'bible'. Wilensky (1986) is a good introduction. Winston and Horn (1984) take an artificial intelligence perspective.

Abelson and Sussman (1985) is the standard reference for Scheme and is based around an introductory computer science course. Dybvig (1987) is a more traditional programming language text.

References

Abelson, H., Sussman, G. J. and Sussman, J. 1985. *Structure and Interpretation of Computer Programs*, Cambridge, MA: MIT Press

Backhouse, R. C. 1986. *Program Construction and Verification*, Englewood Cliffs, NJ: Prentice-Hall

Backus, J. W. August 1978. 'Can programming be liberated from the von Neumann style? A functional style and its algebra of programs' *Communications of the ACM*, **21**(8), 613–41

Bailey, R. 1987. An introduction to Hope. In *Functional Programming: Languages, Tools and Architectures* (Eisenbach, S., ed.), Chichester: Ellis Horwood

Barendregt, H. P. 1981. *The Lambda Calculus: Its Syntax and Semantics*, Amsterdam: North-Holland

Bird, R. and Wadler, P. 1988. *Introduction to Functional Programming*, New York: Prentice-Hall

Bornat, R. 1987. *Programming from First Principles*, Englewood Cliffs, NJ: Prentice-Hall

Boutel, B. 1987. Combinators as machine code for implementing functional languages. In *Functional Programming: Languages, Tools and Architectures* (Eisenbach, S., ed.), Chichester: Ellis Horwood

Brady, J. M. 1977. *The Theory of Computer Science*, London: Chapman and Hall

Burge, W. H. 1975. *Recursive Programming Techniques*, Reading, MA: Addison-Wesley

Burstall, R. M., Collins, J. S. and Popplestone, R. J. 1977. *Programming in POP-2*, Edinburgh: Edinburgh University Press

Cardelli, L. 1983. 'The functional abstract machine' *Polymorphism*, **1**(1)

Carrick, R., Cole, J. and Morrison, R. 1986. *An Introduction to PS-Algol Programming*, PPRR 31, Department of Computational Science, University of St Andrews, Scotland

Church, A. 1941. *The Calculi of Lambda Conversion*, Princeton, NJ: Princeton University Press

Clocksin, W. F. and Mellish, C. S. 1981. *Programming in Prolog*, Berlin: Springer-Verlag

Cohen, B., Harwood, W. T. and Jackson, M. I. 1986. *The Specification of Complex Systems*, Wokingham: Addison-Wesley

Cope, T. 1981. *Computing using BASIC: an interactive approach*, Chichester: Ellis Horwood

Cripps, M., Field, T. and Reeve, M. 1987. An introduction to ALICE: a multiprocessor graph reduction machine. In *Functional Programming: Languages, Tools and Architectures* (Eisenbach, S., ed.), Chichester: Ellis Horwood

Dybvig, R. K. 1987. *The SCHEME Programming Languages*, Englewood Cliffs, NJ: Prentice-Hall

Field, A. J. and Harrison, P. G. 1988. *Functional Programming*, Wokingham: Addison-Wesley

Findlay, W. and Watt, D. A. 1981. *Pascal: An Introduction to Methodical Programming*, London: Pitman

Gehani, N. and McGettrick, A. D. 1986. *Software Specification Techniques*, Wokingham: Addison-Wesley

Glaser, H., Hankin, C. and Till, D. 1984. *Principles of Functional Programming*, Englewood Cliffs, NJ: Prentice-Hall

Gordon, M. J. C. 1979. *The Denotational Description of Programming Languages: An Introduction*, New York: Springer-Verlag

Gries, D. 1981. *The Science of Programming*, New York: Springer-Verlag

Hankin, C., Till, D. and Glaser, H. 1987. Applicative languages and data flow. In *Functional Programming: Languages, Tools and Architectures* (Eisenbach, S., ed.), Chichester: Ellis Horwood

Harper, R., MacQueen, D. and Milner, R. March 1986. *Standard ML*, ECS-LFCS-86-2, LFCS, Department of Computer Science, University of Edinburgh, Scotland

Harper, R., Milner, R. and Tofte, M. August 1987. *The Semantics of Standard ML*, Version 1, ECS-LFCS-87-36, LFCS, Department of Computer Science, University of Edinburgh, Scotland

Harrison, P. and Khoshnevisan, H. 1987. A functional algebra and its application to program transformation. In *Functional Programming: Languages, Tools and Architectures* (Eisenbach, S., ed.), Chichester: Ellis Horwood

Harrison, P. and Khoshnevisan, H. 1987. An introduction to FP and the FP style of programming. In *Functional Programming: Languages, Tools and Architectures* (Eisenbach, S., ed.), Chichester: Ellis Horwood

Hayes, I. 1987. *Specification Case Studies*, Englewood Cliffs, NJ: Prentice-Hall

Henderson, P. 1980. *Functional Programming: Application and Implementation*, Englewood Cliffs, NJ: Prentice-Hall

Henson, M. C. 1987. *Elements of Functional Languages*, Oxford: Blackwell

Hindley, J. R. and Seldin, J. P. 1987. *Introduction to Combinators and λ Calculus*, Cambridge: Cambridge University Press

Jones, C. B. 1986. *Systematic Software Development Using VDM*, Englewood Cliffs, NJ: Prentice-Hall

Kernighan, B. W. and Ritchie, D. M. 1978. *The C Programming Language*, Englewood Cliffs, NJ: Prentice-Hall

Kleene, S. C. 1952. *Introduction to Metamathematics*, Amsterdam: North-Holland

Manna, Z. 1974. *Mathematical Theory of Computation*, New York: McGraw-Hall

Mendelson, E. 1964. *Introduction to Mathematical Logic*, New York: Van Nostrand Reinhold

Minsky, M. 1972. *Computation: Finite and Infinite Machines*, London: Prentice-Hall

Naur, P. *et al.* 1963. 'Revised Report on the Algorithmic Language ALGOL 60' *Communications of the ACM*, **6**(1)

Pagan, F. G. 1976. *A Practical Guide to Algol 68*, London: Wiley

Peter, R. 1967. *Recursive Functions*, New York: Academic Press

Peyton-Jones, S. L. 1987. *The Implementation of Functional Programming Languages*, Englewood Cliffs, NJ: Prentice-Hall

Queinnec, C. 1983. *LISP*, London: Macmillan

Rayward-Smith, V. J. 1986. *A First Course in Computability*, Oxford: Blackwell

Revesz, G. 1988. *Lambda-Calculus, Combinators and Functional Programming*, Cambridge: Cambridge University Press

Richards, M. and Whitby-Strevens, C. 1982. *BCPL – The Language and its Compiler*, Cambridge: Cambridge University Press

Sadler, C. and Eisenbach, S. 1987. Why Functional Programming? In *Functional Programming: Languages, Tools and Architectures* (Eisenbach, S., ed.), Chichester: Ellis Horwood

Schmidt, D. A. 1986. *Denotational Semantics: A Methodology for Language Development*, Boston: Allyn and Bacon

Shapiro, S. C. 1986. *LISP: An Interactive Approach*, Rockville, MD: Computer Science Press

Steele Jr, G. L. 1984. *Common LISP: The Language*, Digital

Stoy, J. E. 1977. *Denotational Semantics: The Scott-Strachey Approach to Programming Language Theory*, Cambridge, MA: MIT Press

Strachey, C. 1967. Fundamental concepts in programming languages. In *Notes for the International Summer School in Computer Programming*, Copenhagen

Turner, D. 1987. An introduction to Miranda. In *The Implementation of Functional Programming Languages* (Peyton-Jones, S. L.), Englewood Cliffs, NJ: Prentice-Hall

Turner, D. December 1976. *SASL Language Manual*, St Andrews, Scotland: Department of Computational Science, University of St Andrews

Turski, W. M. and Maibaum, T. S. E. 1987. *The Specification of Computer Programs*, Wokingham: Addison-Wesley

Wegner, P. 1971. *Programming Languages, Information Structures and Machine Organisation*, London: McGraw-Hill

Wikstrom, A. 1987. *Functional Programming Using Standard ML*, London: Prentice-Hall

Wilensky, R. 1986. *Common LISPcraft*, New York: Norton

Winston, P. H. and Horn, B. K. P. 1984. *LISP*, Reading, MA: Addison-Wesley

Index

In this index, the main references are indicated in **bold**.